Underwater Intervention 2017

New Orleans, Louisiana, USA
21 - 23 February 2017

ISBN: 978-1-5108-4014-0

Printed from e-media with permission by:

Curran Associates, Inc.
57 Morehouse Lane
Red Hook, NY 12571

Some format issues inherent in the e-media version may also appear in this print version.

Printed by Curran Associates, Inc. (2017)

For permission requests, please contact Underwater Intervention
at the address below.

Underwater Intervention
5206 Cypress Creek Parkway
Suite 202
Houston, Texas 77069
USA

Phone: 281-893-8539

www.underwaterintervention.com

Additional copies of this publication are available from:

Curran Associates, Inc.
57 Morehouse Lane
Red Hook, NY 12571 USA
Phone: 845-758-0400
Fax: 845-758-2633
Email: curran@proceedings.com
Web: www.proceedings.com

TABLE OF CONTENTS

ROV SESSIONS

SUPPLIER SOLUTIONS

RINI TECHNOLOGIES

Free-swimming Diver Heating System (FDHS)

22 February 2017

Underwater Intervention

Dr. Dan Rini – President & CEO

RINI Technologies, Inc.
582 S. Econ Circle
Oviedo, FL. 32765
(407) 359-7138
info@rinitech.com

Diver Thermal Protection Need

- Active heating needed for long duration, cold water dives
- Two types of viable commercially available systems:
  1) Large Surface Supply Heaters
     - Supply ample amount of heat
     - Open loop – hot water suit
     - Infrastructure support
     - Large umbilical
  2) Portable Electric Resistive Systems
     - Inadequate heating
     - Prone to damage
     - Poor fit can burn diver's skin

RINI TECHNOLOGIES

2

RINI Approach

- Thermodynamic heat pump
- Supply meaningful amount of heat
- Heat source close to diver
  - Reduce heat losses
  - Minimize infrastructure support
- Use benign heat transfer medium

RINI TECHNOLOGIES

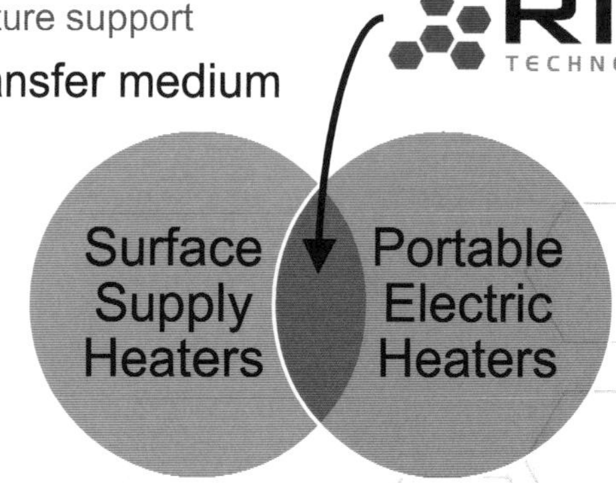

RINI TECHNOLOGIES

3

RINI Tubesuit

- **Heat** delivered via **"tubesuit"**
  - Closed loop circulates warm water
    - Eliminates hot spots
  - Full body heating

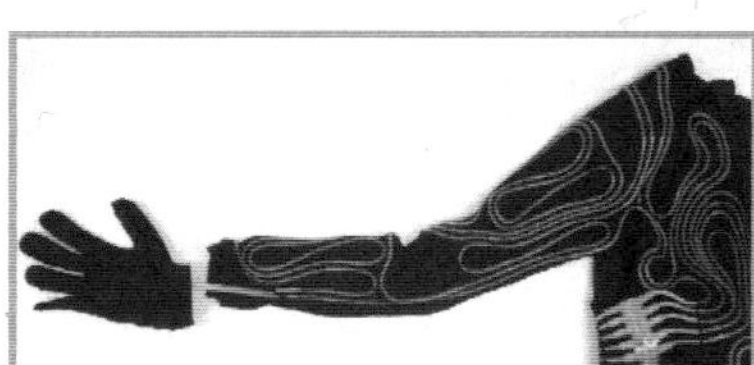

RINI TECHNOLOGIES

4

RINI Diver Heating Unit

- Technology: **Thermodynamic Heat Pump**
  - **100 W electrical power ≈ 300 W heat**
  - Extracts heat from cold seawater
- Electrically powered
  - Operates from Diver Power Pack or surface power
- Developed in cooperation with U.S. Navy
  - Authorized for Military Use

RINI

5

RINI Diver Power Pack

- Uses 2 Military Batteries
  - High capacity rechargeable
- Wet-mate Connector
- On/Off Switch
- Built to IMCA & Mil. Standards
- Bracket Hard Mounts to #80 Tank

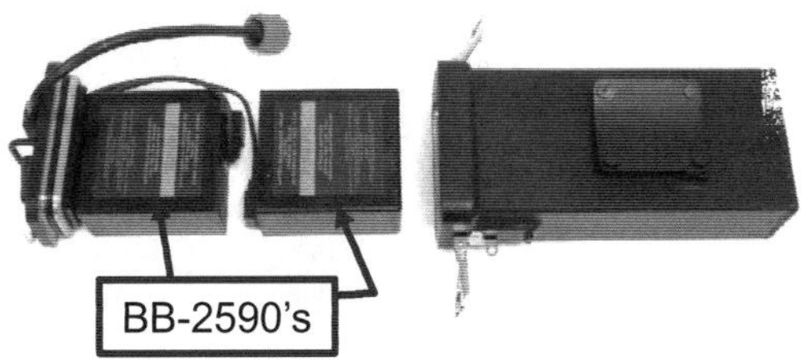

RINI

6

Free-swimming Diver Heating System (FDHS) Components

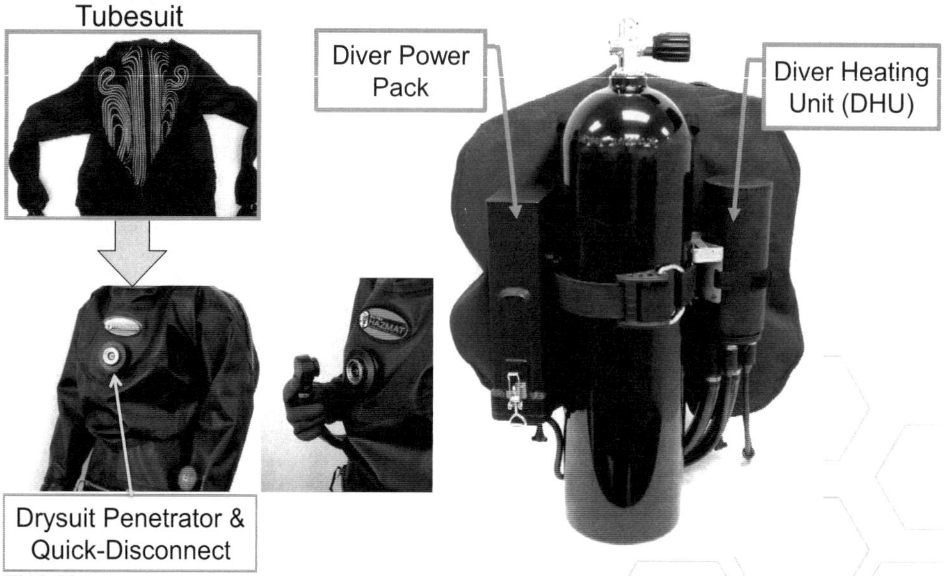

RINI TECHNOLOGIES

7

Current Use

- Lab Testing:
  - RINI
  - U.S. Navy Experimental Dive Unit
- Demonstrate thermal performance
- Support AMU Technical Data Package
- Month long training dives
  - Daily 6 hour dives to depth
  - 35°F to 40°F water
  - Body core & digit temperature monitored
    - Satisfactory throughout duration

RINI TECHNOLOGIES

8

RINI TECHNOLOGIES

Free-swimming Diver Heating System (FDHS)

Dr. Dan Rini – President & CEO

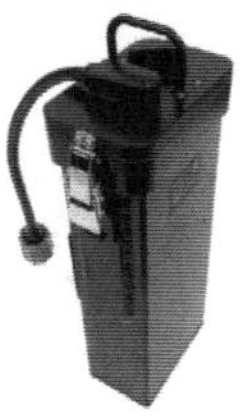

RINI Technologies, Inc.
582 S. Econ Circle
Oviedo, FL. 32765
(407) 359-7138
info@rinitech.com

VALVES AND CYLINDER SAFETY

Avoiding dangerous failures

INJURIES OCCUR DUE TO POORLY MAINTAINED CYLINDERS

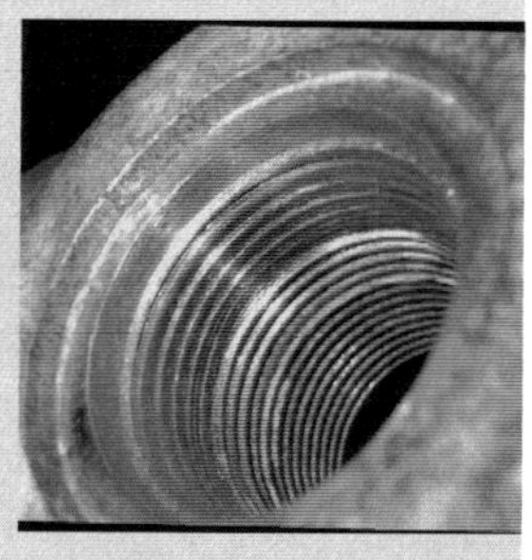

Corrosion

Cracks

External Damage

Valve / Threads

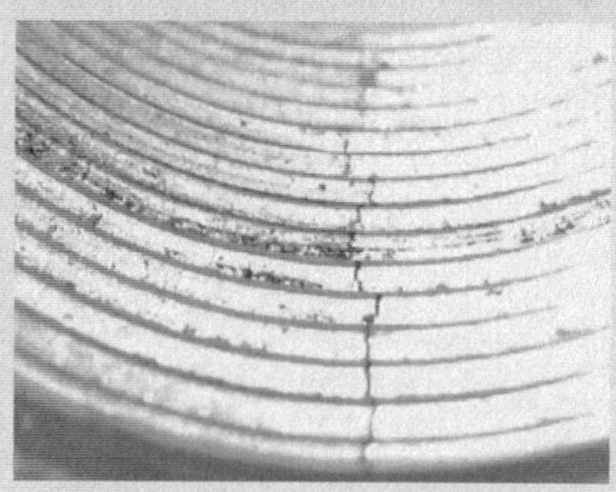

MOISTURE CAUSES CORROSION

- This Bailout bottle had approximately 1 cup (8 ounces) of water
- Corrosion quickly develops – creating a risk of rupture

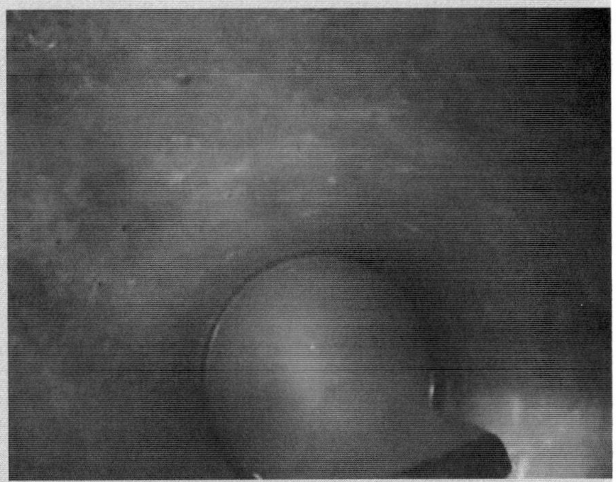

NON-COMPATIBLE VALVE PLACED IN INCORRECT CYLINDER

The International Marine Contractors Association (IMCA) reported at least 4 incidents

Valves of incorrect size placed within cylinders causing extreme damage to valves

1 Injuries due to Failure of Diver's Emergency Gas Cylinder

A member has reported an incident which a valve failed on a diver's emergency cylinder. The incident occurred on board a diving support vessel (DSV) whilst divers were preparing for a dive and were putting on their diving suits. A valve forcefully parted from a high pressure (HP) compressed air cylinder filled to 180 bar. Five divers were injured by the parted valve as it flew off the cylinder.

1 Pillar Valve Failure

A member has reported an incident in which a pillar valve from an emergency gas cylinder separated from the bottle while under pressure, rapidly releasing 50 cu ft of air to atmosphere and hitting a diver's helmet. A trial run dive was planned in an anchorage prior to mobilisation to a job site. Dive checks were carried out as per standard procedures and the diver was cleared to enter the dive basket, ready for the water. As the diver turned upon entering the basket a loud bang was heard and there followed an escape of gas. The diver collapsed to the stage grating and suffered multiple contusions and a dental insult. He was evacuated to shore for further evaluation and treatment but made a full recovery.

3 Diver Injury During Air Cylinder Recharging

A member has reported a blown-out pillar valve resulting in injury to a diver working on deck.

2 Near-Miss Involving Bail-Out Bottle Pillar Valve

Keywords: Pressure

After replacement of the pillar valve on a bail-out bottle, the bottle was being filled with air. When the pressure reached about 100 bar, the threads slipped from the bottle and the pillar valve flew out. Fortunately, nobody was injured and nothing was damaged.

THE PRIMARY CONCERN

Cylinder manufacturers don't make valves and Valve Manufacturers don't make cylinders

No Unique Markings

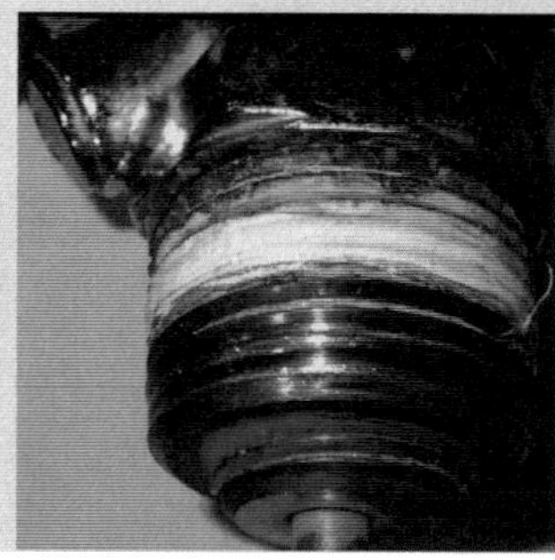

MARCH 2004 – NO INJURY

- Valve replaced because hand wheel was slipping (poor valve maintenance)
- Experienced diver located spare valve
- Incorrect valve for the cylinder
- At 100 BAR (1450psi)the valve came out

AUGUST 2009 – HIT HELMET / NON SERIOUS INJURY TO DIVER WHO WAS WEARING CYLINDER

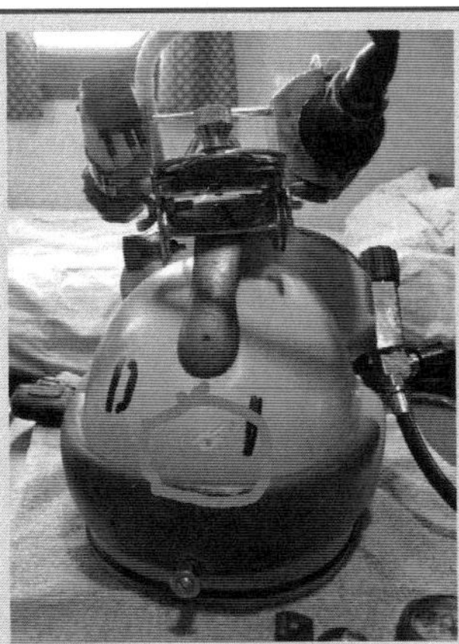

- Diver checks completed
- Diver cleared to enter the basket
- Diver turned and loud bang heard
- Gas escaped and diver collapsed
- Diver evacuated
- Cylinder hydrostatically tested off site and returned
- Cylinder had been charged for several days prior to the incident
- Wrong valve inserted in cylinder – damaged the threads

JULY 2010 – INJURED RIGHT EYE

- Diver re-charging cylinders after being delivered empty
- Members of dive team heard loud noise
- Injured diver found on ground near re-charging station
- Valve disconnected from cylinder
- Inured eye of fill station technician caused by filling hose of the air compressor

OCTOBER 2014 – FIVE DIVERS INJURED

- 5 divers preparing for a dive on a dive support vessel
- At 180 BAR (2610psi)the valve parted from the cylinder
- 5 divers injured
- Incompatible valve placed in cylinder

THREAD SIZING

- These parallel threads are very similar but not compatible, as pitch, pitch diameter and thread forms may be different.
    - BS 3643 (25mm) vs ISO 228 (G3/4-inch)
- Greater torque is required to insert an M25X2 pillar valve into a ¾ inch parallel thread after a few rotations.

PROPER TORQUE

Enough torque to contain the gas

- Pillar valves should be inserted within a cylinder without the use of any tools
  - Exception with Tapered thread valves
- A proper working O-Ring contains the pressurized gas within the cylinder with a minimal amount of torque
- The torque is applied after inserting the valve and should only be enough torque to contain the gas (70-90ft lbs)
- Cylinder manufacturers and valve manufacturers may give differing torque values

DAMAGED THREADS

Excessive torque damages cylinder threads

If a cylinder thread is damaged or valve threads are damaged

DO NOT USE

CURRENT SAFETY CONCERNS

- Receiving cylinder with valve in place
- Measure valves and cylinder
- Placing valve in cylinder
- Marking valve / Marking Cylinder
- Inspection

HOW TO ENSURE THE VALVE IS CORRECT IN A CYLINDER BROUGHT TO THE SITE WITH THE VALVE INSERTED

- If valve is inserted off site
  - there is no way to verify correct valve within the proper cylinder
- There is no standard valve markings for specific cylinders
- The valve must be removed and checked by trusted / trained staff

MEASURING THREADS

SOLUTION

- You can Measure the threads
- Ensure the exterior of the valve matches the interior of the cylinder threads

PROBLEM

- Measuring device on hand
- How do you know it was checked by a qualified person

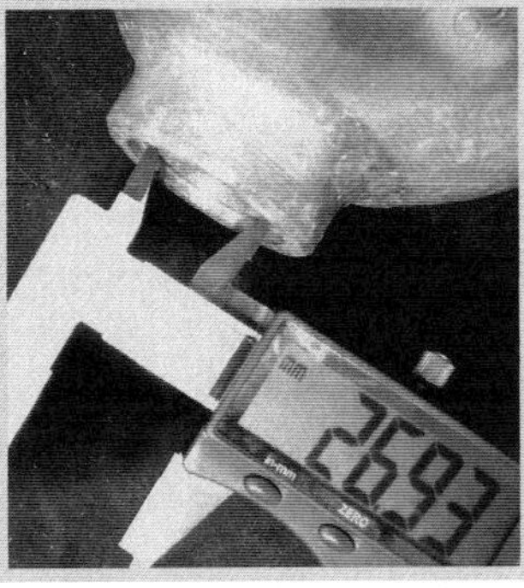

CYLINDER AND VALVE ARE SEPARATED WHEN THEY ARRIVE ON SITE

- If the cylinder arrives with no valve
  - Ensure no contaminates entered cylinder
- Proper training for persons responsible for checking the valve, interior of cylinder and cylinder threads for any damage
- Certain that the valve being inserted is the proper valve
  - Proper training

SPECIFIC MARKINGS ON CYLINDERS AND VALVES

- Having a system in place to ensure that only the proper valves are located on site with the proper cylinders
  - Valves and cylinders marked prior to arrival
- How do you verify that the markings were correct before arrival?
- Who conducted the verification?
- What markings are used? What do they mean?
- How do you verify off site facilities
  - Hydrotesters?
- Company wide system must be ensured and followed to gain confidence of on site staff

SUGGESTED BEST SOLUTION

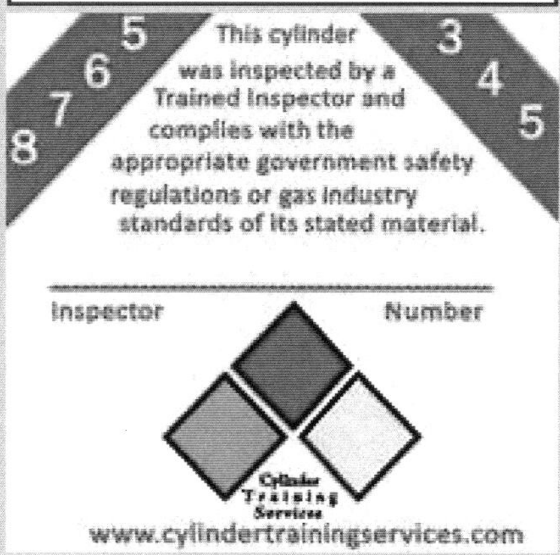

- All cylinders coming on site are empty and devalved
- Cylinders inspected for damage
  - Internal / external and threads
- Valves inspected for damage
- New O-rings used on valve
- Valves inserted – hand tight – then torqued
- Paint marking on valve and cylinder
  - Security mark
- Inspection sticker signed by authorized personnel
  - approved by safety officer
  - Current date

NO INDUSTRY STANDARD

Since there is no industry standard on properly checking the cylinder and ensuring the proper valve is inserted it is recommended that each location / site or company develop their own written safety protocols

PROPERLY TRAINED PERSONNEL

- Companies should have written inspection protocols or follow a specific program
- The primary point of concern is ensuring that personnel are trained in the inspection of cylinders.
- Personnel must follow a standard inspection criteria
- Everyone must understand the basic markings on cylinders
- Personnel must know to look for safety mark on cylinder / valve to feel comfortable that the proper valve is inserted into the proper cylinder

Development of an Enhanced Underwater Navigation Capability for Military Combat Divers

Dennis G. Gallagher, Richard J. Manley, William W. Hughes III, Allie M. Pilcher
Naval Surface Warfare Center – Panama City Division
Underwater Systems Development and Acquisition Branch
Panama City, FL USA 32407

ABSTRACT

Military combat divers conduct underwater navigation missions in conditions with extremely poor visibility. Critical to the success of these missions is accurate underwater navigation, as well as staying within closed circuit oxygen rebreather depth limits. Unfortunately, handheld or tactical board mounted compasses, depth gauges, and chronometers can become virtually useless in the zero visibility environment. This has been an ongoing and serious limitation to military combat diving operations.

Naval Surface Warfare Center-Panama City Division (NSWC PCD) is the U.S. Navy's leading laboratory for research, development, testing, evaluation, and technology transition of diving and life support systems, including diver visual display systems. Under a recent joint government development program, NSWC PCD developed a prototype enhanced underwater navigation system for military combat divers. The prototype comprises a micro display, optical lens, electronic compass, depth sensor, microprocessor, associated electronics, and battery integrated into an adjustable position optical display that is mounted to the side of a low volume dive mask. The system provides clear visual display of the diver's compass heading, depth, and swim time regardless of poor visibility conditions.

Divers can turn the system ON/OFF, adjust display brightness, use the stopwatch function, and lock their compass heading using the NAVLOK feature. When the NAVLOK button is pressed, the displayed compass heading is boxed, and a trackball icon appears to assist divers in staying on heading. The system operates on a single 3V lithium ion battery which provides 20 + hours of operation.

Joint service military divers evaluated the prototypes, conducting a series of underwater navigation exercises in both daylight and nighttime conditions. Participants strongly recommended rapid transition of this new underwater navigation tool to military production status.

This paper describes the technical and operational need for such a system by the combat diving community, specifics of the prototype design phases, results of in-water testing by joint service combat divers, and the rapid technology transition plan using a Cooperative Research And Development Agreement (CRADA) for bringing this much needed capability to the joint service military diving community.

Keywords: *head-up display, diving, military, underwater navigation*

INTRODUCTION

Military combat divers conduct underwater navigation missions in conditions with extremely poor visibility. Critical to the success of these missions is accurate underwater navigation, as well as staying within closed circuit oxygen rebreather depth limits. Unfortunately, handheld or tactical (TAC) board (Figure 1) mounted compasses, depth gauges, and chronometers can become virtually useless in the zero visibility environment. As a r esult, there is an ongoing technical and operational need for an enhanced underwater navigation tool that can address these limitations.

Figure 1 – Standard TAC-boards used for underwater navigation

The Naval Surface Warfare Center-Panama City Division (NSWC PCD) is the U.S. Navy's leading laboratory for research, development, testing, evaluation, and rapid technology transition of diving and life support systems, including diver visual display systems. Under a recent joint government program, NSWC PCD developed a prototype enhanced visual display underwater navigation system for military combat divers to address this technical and operational need.

COMBAT SWIMMER NAVIGATION MASK (CSNM)

Under the first phase of the joint government program, NSWC PCD developed a prototype Combat Swimmer Navigation Mask (CSNM) based on the standard U. S. Navy Rescue Diver mask. The prototype was designed to provide a clear, visual display of the diver's magnetic compass heading, depth, and swim time regardless of poor visibility conditions. A high resolution, active matrix organic light emitting diode (AMOLED) micro display was used with a fixed position optical lens in the lower right corner of the mask. Electronic compass, depth sensor, microprocessor, associated electronics, battery, and control buttons were built into the left and right side of the mask frame. Additional miniature light emitting diodes (LEDs) were integrated inside the lower left and right side corners of the mask to provide additional navigation cues. (Figure 2)

DISTRIBUTION STATEMENT A: Approved for public release; distribution is unlimited.

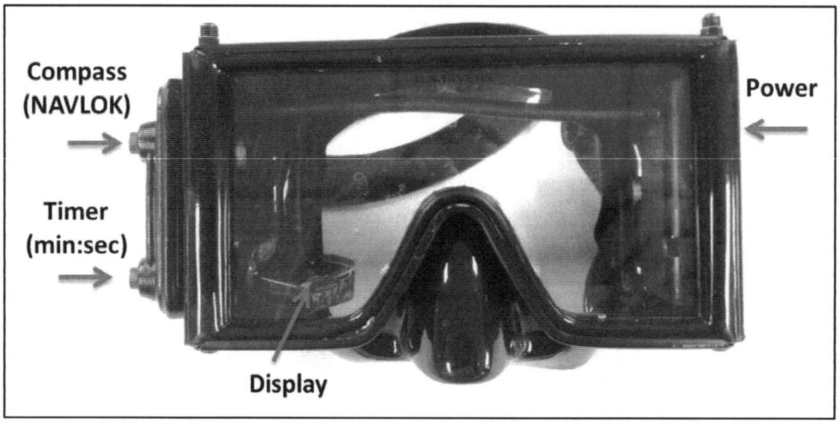

Figure 2 – CSNM Prototype

The display data layout is shown in Figure 3. This includes a digital compass display on top (in degrees), the diver's depth (in feet) in the middle line, and the swim time/timer in the bottom line (minutes:seconds). The lower left side of the screen includes a remaining battery life icon. Over the compass heading display are additional navigation icons.

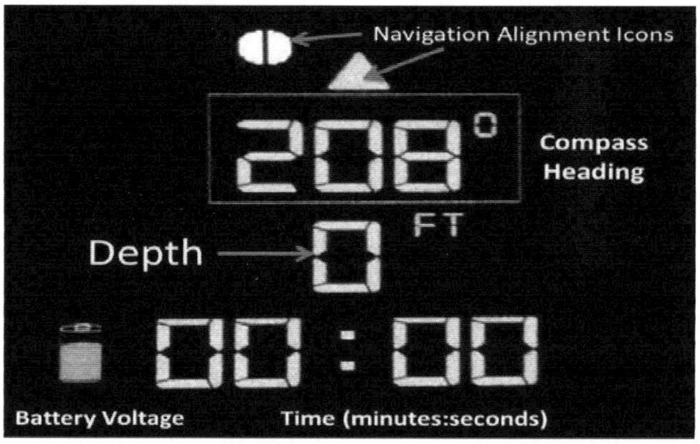

Figure 3 – CSNM Display Screen Layout

The diver is able to turn the system ON/OFF, adjust display brightness levels, start/stop/reset the time, calibrate the compass, and navigate using the NAVLOK function. When the NAVLOK button is pressed, the selected compass heading is "boxed", navigation/alignment icons are displayed, and the LEDs are illuminated. The diver can use the boxed compass display, alignment icons, and LEDs which alternately blink when the diver is +/- 5 degrees or more off his selected heading to assist in underwater navigation. The diver can also turn the NAVLOK function off, and then reset it as often as required.

End User Evaluations

Joint service military divers evaluated the CSNM prototype using the U. S. Navy MK-25 (Drager LAR 5) in standard 2-man combat navigation teams (Figure 4).

DISTRIBUTION STATEMENT A: Approved for public release; distribution is unlimited.

Figure 4 – Combat Divers during CSNM day evaluations

Each team conducted a series of underwater combat navigation exercises performed in daylight and night conditions (Figure 5).

Figure 5 – Combat Divers during CSNM night evaluations

At the conclusion of the evaluation each diver was asked a series of questions from a Performance Evaluation Questionnaire (PEQ), providing feedback regarding functional and operational requirements, system design, human factors interoperability, and changes required for continuing development of the enhanced underwater combat navigation capability.

Divers determined that the system should be redesigned as an external module mounted to a low volume dive mask with a flip-down/flip-up display, instead of integrated inside of a dive mask. Additional recommendations were to eliminate the LEDs, and to make the flip-down optical module align directly in front of the eye instead of in the upper, lower, or side position. Also, to reduce cost and eventually make the finished version affordable for dive commands, a low power, low cost micro display should be investigated. From the end users' input, NSWC PCD developed the Combat Diver Navigation Module (CDNM).

COMBAT DIVER NAVIGATION MODULE (CDNM)

In the second phase of the joint government program, NSWC PCD developed the prototype Combat Diver Navigation Module (CDNM) based on feedback from joint service divers' evaluation of the previous CSNM system. This prototype was designed to provide the same clear, visual display of the diver's magnetic compass heading, depth, and swim time regardless of poor visibility conditions. The CDNM comprises a lower power passive matrix organic light emitting diode (PMOLED) micro display, plano-convex magnifying lens with red lens filter, upgraded 6-axis electronic compass, depth sensor, microprocessor, associated electronics, and battery integrated into an external housing mounted to the right side of a low volume dive mask with an adjustable position display. The 3-volt lithium battery provides 20+ hours of operation. The diver is able to turn the system ON/OFF, adjust display brightness levels, start/stop/reset the time, calibrate the compass, and navigate using the NAVLOK function (Figure 6).

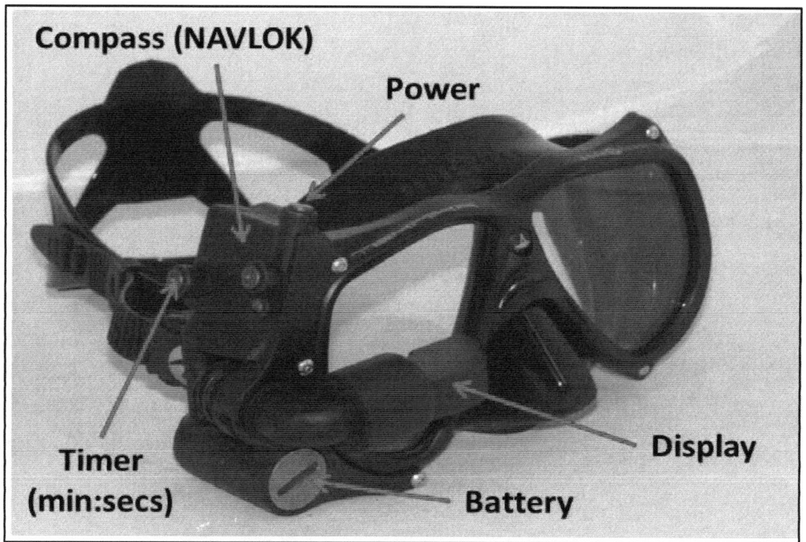

Figure 6 – CDNM Prototype

The display layout provides the magnetic compass heading with larger numbers at the top of the display with an alignment mark above the compass reading. The lower right side can display diver depth in either feet or meters, the lower left side displays time in minutes and seconds, and the lower middle section displays the remaining battery life icon. The prototypes used red lens filters to provide low light red characters (Figure 7) to reduce light signature and preserve operator night vision.

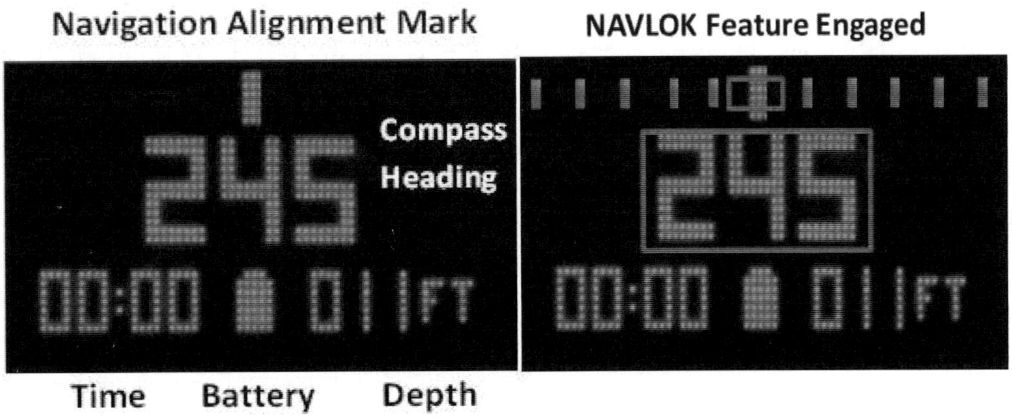

Figure 7 – CDNM display screen layout

End User Evaluations

Joint service military divers conducted a series of expanded end-user evaluations using the CDNM prototype, U. S. Navy MK-25 (Draeger LAR V), and standard underwater navigation TAC-board in standard 2-man combat navigation teams (Figure 8). The evaluations comprised a series of underwater combat navigation exercises performed in daylight and night conditions which included performing a dog-leg underwater navigation swim.

Figure 8 – Combat Diver during CDNM Evaluations

Each team would take a heading toward a target, submerge, and navigate to the target. Upon reaching the target the divers would reset the NAVLOK on a heading toward a second target, submerge and repeat this exercise. Upon reaching the last target divers would repeat the process during the return leg, switching roles as lead navigator.

DISTRIBUTION STATEMENT A: Approved for public release; distribution is unlimited.

Divers conducted the exercises using a standard TAC-board for navigation, depth monitoring, and time; and then repeated the exercise using the CDNM prototype for navigation, depth monitoring, and time.

At the conclusion of the evaluations each diver was asked a series of questions from a Performance Evaluation Questionnaire (PEQ), providing feedback regarding final functional and operational requirements, system design, human factors interoperability, and any changes required for final development of a military production version CDNM.

One hundred percent (100%) of joint service divers participating in the user evaluations rated the CDNM as Good or Excellent in overall performance.

Recommendations

As a result of the successful prototype development and test results, the Supervisor of Diving (NAVSEA 00C3) briefed the CDNM to the DoD Military Technical Acceptance Board (MTAB). Representatives from the Navy, Marine Corps, Army, and Air Force subsequently voted the CDNM as a **Common Type** diving requirement for the joint military services.

Based on the positive feedback from the end users and the MTAB designation, NSWC PCD proposed a follow-on rapid *transition-to-production* project. This phase would use a Cooperative Research And Development Agreement (CRADA) with a commercial industry partner to accelerate final development, testing, and production status.

CURRENT STATUS

NSWC PCD has entered into a Cooperative Research And Development Agreement (CRADA) with an industry partner to jointly develop a military production version CDNM, manufacture initial pre-production units, and conduct end user testing. Anticipated completion is Fall 2017.

DISTRIBUTION STATEMENT A: Approved for public release; distribution is unlimited.

Improving Coiled Tubing Safety With An Autopark Brake

Benton F. Baugh, Ph.D., P.E.[1], Eston Huang[2], and James R. Crawford[3]

Abstract— **A discussion of using non-marking slips on Coiled Tubing units to automatically stop the Coiled Tubing in the case of a runaway situation, and in normal operations offering a substantial safety upgrade in both situations.**

I. INTRODUCTION

When a string of Coiled Tubing or other type of umbilical are suspended in deep offshore water from a device on the surface, a chance always exists of the device losing its grip. In the case of a reel or Coiled Tubing Injector Head the motor and/or brakes can fail. The Injector Heads also have the potential of simply losing the grip on the Coiled Tubing. In either case it can be a catastrophic event, both to equipment and personnel. This paper discusses a method of automatically sensing when the Coiled Tubing or umbilical starts slipping in the Injector Head or exceeds a predetermined speed in the case of a reel.

II. GLOSSARY

Automatic Parking Brake: A braking mechanism which automatically sets during specified situations.

BOP: Blowout preventer.

Coiled Tubing: A thin walled pipe rolled up on a reel for injecting into and oil / gas well or a pipeline for service operations.

Injector Head: An assembly with opposing chains which engage a Coiled Tubing to run and retrieve it.

Pressure control equipment: BOPs or Strippers to contain wellbore or pipeline pressures.

Slips: devices typically for supporting pipe in a well. They will characteristically have sharp teeth engaging the pipe and an 8 degree taper engaging a bowl so that as the slips slide down the bowl they more tightly engage the pipe.

Snubber: A device for pushing a tubular into and pulling it out of a pressurized environment.

[1]Benton F. Baugh, Ph.D., P.E. is a principal of Baugh Consulting Engineers, Inc. and adjunct professor at the University of Houston, bbaugh@baughengrs.com.

[2]Eston Huang is a design engineer for ReelPower Oil and Gas, Inc. Houston, ehaung@reelpower.com

[2]James R. Crawford is a principal at Crawford Technical Services, Inc., jimbob@crawfordtechnical-serv.com.

Snubber Stand: A unit for mounting and Automatic Parking Brake below a Snubber.

Synthetic Rope: a rope manufactured of non-steel product such as a polymer.

Umbilical: An assembly of multiple inner lines for doing subsea service operations. The outer layer can be spiral wrap steel wire a plastic type material. The inner layers can be hydraulic, electrical, or fiber optic.

III. DISCUSSION

Fig. 1 shows a rig floor after a runaway Coiled Tubing situation in which substantial damage was done around the rig floor. Fig. 2 shows a similar situation when a reel was deploying and Umbilical and it became a runaway situation. The figure is actually a screen shot of a movie shot during the incident. Shortly after this part of the movie, the entire reel came out of its stand and moved towards the rear of the vessel. Clearly these incidents indicate a need for a better way to stop a dangerous runaway situation.

This paper discusses a method for not only stopping a runaway situation, but also automatically sensing the situation and stopping the Coiled Tubing or Umbilical at the earliest possible time. A runaway has a sort of “freight train” characteristic in that the longer you wait to stop it, the faster it starts to accelerate.

Fig. 3 shows an Automatic Parking Brake placed below a Coiled Tubing Injector Head. If the Coiled Tubing or Umbilical is being lowered into the seawater from a vessel, Automatic Parking Brake may be the lowest piece of equipment above the supports. If the Coiled Tubing or Umbilical are being lowered directly into a riser or wellhead, pressure control equipment would be between the Automatic Parking Brake and the pressure control equipment.

Fig. 4 shows an Automatic Parking Brake in a Snubber Stand. The Snubber Stand must withstand the forces of the coiled tubing or snubbing operations, and must have appropriate mounting profiles on the upper and lower plates.

Fig. 5 Shows a small control panel with a rotary valve to manually open and close the Automatic Parking Brake and an E-Stop Button which can be hit in an emergency to quickly close the Automatic Parking Brake. This small panel is only an illustration of the controls which would be integrated into the normal operators control panels.

Fig. 6 shows an Automatic Parking Brake mounted ???? feet above the floor and with 1.315" outside diameter pipe being dropped. A roller wheel on the Coiled Tubing Sensor Assembly which is spring loaded against the pipe senses when the pipe is dropped. The controls compare the speed of the dropping pipe with the speed of the Injector Head Sensor (which is not moving), detects a discrepancy, and commands the Automatic Parking Brake to stop. The results of this testing were:

trial #	dist (in)	time (sec)
1	33	0.41
2	31.5	0.40
3	30.5	0.40
4	42	0.47
5	30.5	0.40
6	43.5	0.47

Fig. 7 The Coiled Tubing Sensor has a wheel which is spring loaded against the coiled tubing (Or Umblical) and gives a rotational reading which is a function of the Coiled Tubing or Umbilical Speed.

Fig. 8 The Injector Head Sensor is mounted on the rear of one of the motors in the Coiled Tubing Injector head, with an output converted to coordinate with the Coiled Tubing Sensor Assembly when everything is operating satisfactorily.

Referring back to Fig. 2 where the Umbilical is being deployed with the power of a reel over a sheave directly into the water, there is no comparable coordination between an Injector Head Sensor and a Coiled Tubing Sensor. In this case and over speed sensing is required. This can be done either by a wheel running on the Umbilical or a sensor measuring the actual reel rotation rate, or with an E-Stop button.

A significant problem with engaging Coiled Tubing or Umbilicals in the past is that "normal" Slips have sharp teeth to increase the friction coefficient against the tubular member to be supported as seen in Fig. 9. This functionally damages the outer surface of the tubular member. This damage is acceptable in large oilfield pipes, but is not acceptable on Coiled Tubing or Umbilicals. For this reason smooth faced slips are utilized for this service.

Fig. 10 shows brass slips being tested on Synthetic Rope which is extremely sensitive to the sharp teeth of conventional slips[4]. The particular synthetic rope in the figure was tested to about 1/3 of its breaking strength. It could not be tested more in this test setup as the synthetic rope is very compressible and the brass slips simply ran out of stroke. The supporting characteristic was "failsafe" (generally meaning that the harder you pull the tighter the grip), so there was no question that higher loads could be pulled if the ability to handle the compression was inproved.

References

[4] UI 2014 Failsafe Slips on Synthetic

Fig. 1 Rig floor after a Coiled Tubing runaway event

Fig. 2 Runaway event from an Umbilical Reel

Fig. 3 Autopark Brake below Injector head and above BOPs

Fig. 4 Automatic Parking Brake in a Snubber stand

Fig. 5 Small control panel with E-Stop Button

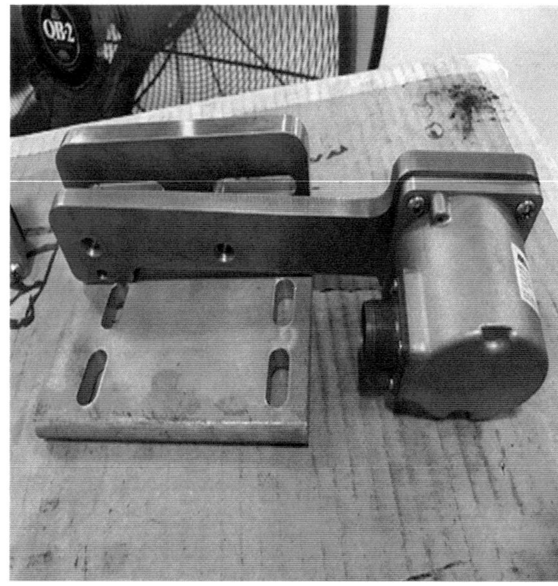

Fig. 7 Coiled Tubing Sensor

Fig. 6 Automatic Parking Brake during time trials

Fig. 8 Injector Head Sensor

Fig. 9 Typical sharp teeth in slips

Fig. 10 Brass slips tested on Synthetic Rope

Synthetic Rope Pipeline Remediation Pigs

Barrett A. Broussard[1], David Pugh[2], Benton F. Baugh, Ph.D., P.E.[3], and James R. Crawford[4]

Abstract— **A discussion of using synthetic rope as a carrier means for pipeline remediation using a jetting pig with venting past the jetting nozzles on the return trip.**

I. INTRODUCTION

Pipelines are in a continual need of cleaning due to wax type buildup occurring on the inner bores. Excessive buildup can lead to reduced production flow and potentially even complete blockages. Additionally, when smart pigs or inspection pigs are sent through the pipelines, the walls are typically required to be cleaned first. The better the cleaning job, the more reliable the inspection pig data will be.

This cleaning can be done by:

1. Free ranging pigs which simply go with the flow and attempt to scrape the buildup off the walls. Results on the methods sometimes indicate that rather than removing the buildup, they compress it onto the wall for temporary advantage while making them harder to remove later.
2. Coiled tubing services where a steel coiled tubing is unrolled into the pipeline as the pig moves forward. This is a relatively expensive method, but as it provides a return path for the flow it is used most often for removing complete blockages.
3. Jetting pig with a hold back feature such as a wire to allow it to move forward but also to provide a restraint to cause a jetting differential pressure across a pig. It typically has the problem that when pulling it back, it must swab the pipeline, or bring the entire pipeline of fluids back with it. When coming back to a platform with a riser, the wire may not have the strength to pull the column of liquid up the riser. The operator potentially has to allow the fluid in the riser column to drain across the jet nozzles of the pig to recover the pig.

II. DISCUSSION

On using the third alternative, providing a synthetic rope instead of a wire line makes the operator capable of having a higher force (e.g. 40,000 lbs.) to allow a high jetting differential across the pig itself. The more robust available force also gives a much greater assurance of rapid recovery.

****

[1]Barrett A. Broussard is a Technical Advisor at BlueFin Services and can be contacted at bbroussard@bluefingrp.com.

[2]David Pugh is President of BlueFin Services and can be contacted at dpugh@bluefingrp.com.

[3]Benton F. Baugh is a principal of Baugh Consulting Engineers, Inc., and adjunct professor at the University of Houston and can be contacted at bbaugh@baughengrs.com.

[4]James R. Crawford is a principal at Crawford Technical Services, Inc. and can be contacted at jimbob@crawfordtechnicalserv.com.

Fig. 1 shows generally what is attempting to be done with a jet pig such as this. It is a photograph of a test setup inside a clear plastic pipe with a multiple port jetting head on the left side of the photo. In this case there are eight jet ports. Across the center of the photograph you can see the jet streams spreading out and hitting the wall on the right side of the photograph. There is a more intense jetting directly when the jet hits the wall, however, the jetting is shown to spread out to the sides such that the impact from one jet hits the impact from the adjacent jet. In this way with a properly specified and tested jet port configuration, there is an assurance that the entire pipe wall is cleaned.

Fig. 2 shows what an assembly looks like to deploy one of these pigs on a synthetic rope. The Synthetic Rope Pig is shown to the right with a synthetic rope attached to an eye at its left side. Immediately to its left is a flange which would connect to the clients pipeline. Past the flange are a Swivel Union, Stripper, a Swivel Union, Stripper, a Swivel Union, and a Synthetic Rope Winch.

The Synthetic Rope Winch is a dual drum type with a motor above each drum. The drums are multi-groove with the Synthetic Rope being wrapped around one and then the other several times. This collective friction, much like a Catahead on a drilling rig, provides a full grip on the Synthetic Rope. The Synthetic Rope exits the Winch Case at its lower right at ninety degrees to the incoming rope and goes off to a light duty storage reel.

An additional feature of this style mounting is that the high tension is contained within the pipeline and winch and is directly in line with the pipeline. This reduces a number of setup complications. The bar sticking out the left side of the winch is on the centerline of the pipeline and allows the unit to be easily supported, keeping the moments off the pipeline connections.

Fig. 3 shows the jetting pattern of an actual Synthetic Rope Pig, which again has eight jetting ports. The number of jetting ports is a function of the material being pumped, the material in the pipeline, the diameter of the pipeline, and anticipated flow rates. The eight jet nozzles in this photograph are 1/8 inch diameter and the flow rate is ???? barrels per minute.

Fig. 4 shows that the jetting has been stopped and in this case the water in the pipeline is simply draining across the nozzles.

Fig. 5 shows that the flow has been restarted, but it is automatically directed away from the jet nozzles to a larger vent port. Each time the flow is stopped and restarted, the flow goes alternately between the jetting ports and the vent port. The application of this is after a pig has moved forward and completed its cleaning task, the pumping is simply stopped. At that time the winch simply pulls the pig back. The speed of pulling the pig back times the cross section area of the bore of the pipeline can be converted to an equivalent gallons per minute or barrels per minute which need to pass through the vent port. Pulling back without pumping is exactly equivalent to not moving the pig and pumping past the pig. Having an internal Selector Valve to do this substantially simplifies the operations and speeds the recovery operations.

Fig. 6 shows the internal workings of a Selector Valve which would be inside the pig. Inside the valve there is a central axle with a Diverter on its left and an Actuator on its right. There is a leaf spring which engages both the Diverter and Actuator. Flow enters from the left and causes the Diverter to stay where it is, but moves the Actuator clockwise to the opposite side wall, exposing the jetting ports. The Actuator has a detent to stay in that position, so when flow stops, the leaf spring moves the Diverter clockwise to align with the new Actuator position. When flow is started again, the Diverter stays in the new position and flow pushes the Actuator back to the original counterclockwise position and directs flow to the vent port. This process happens repeatedly each time flow is stopped and started again. Again, the pulling movement of the pig back up the pipeline thru the pipeline liquids is exactly equal to starting the flow.

Fig. 7 shows a test setup for this type pig. This test setup is 6 inch standard wall pipe, with four 3D bends in place. Most subsea bends are 5D (radius of the bend is 5 times the pipe diameter), so 3D testing is demonstrating an extreme case, and demonstrates flexibility of the system.

Fig. 8 shows a pig with synthetic rope attached after repeated testing, both in jetting and through the pipe test loop. The original pig could not be pulled back through the 3D bends as the extended position of the eye at the top in the photo caused it to bind in the pipelines. In order to be pulled backwards on recovery, brass rollers were added as seen and it pulled backwards without a problem.

The Synthetic Rope Pig System is a viable and economic way to clean and remediate pipelines where there are no entrance restrictions smaller than a 3D bend, and where communication has not been 100% restricted. This tool can be designed to operate in any pipeline environment including gas, crude, or even 3 phase flowlines with the well flowing when utilized with a pressure control system, Hence, this technology has the potential to clean pipelines that are otherwise inaccessible to conventional technologies and so increase their uptime and throughput and the revenues available to their operators.

Fig. 1 Showing jetting pattern diffusion across internal pipe wall

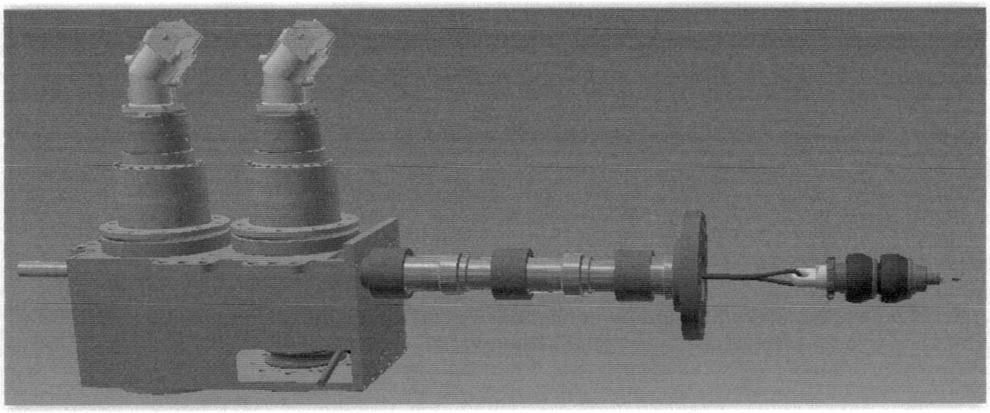

Fig. 2 Synthetic Rope Pig delivery system

Fig. 3 Jetting pattern

Fig. 4 No flow

Fig. 5 Venting pattern

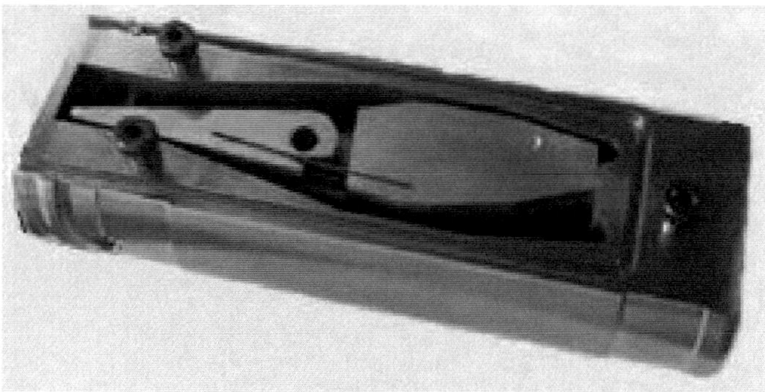

Fig. 6 Selector valve

Fig. 7 Test stand with 3D bends

Fig. 8 Synthetic Rope Pig after testing

SHINKAI 6500 Update

Masanobu Yanagitani

Ship operations and Engineering Group, Research Fleet Department
Marine Technology and Engineering Center (MARITECH)
Japan Agency for Marine-Earth Science and Technology (JAMSTEC)
2-15, Natsushima-cho, Yokosuka, 237-0061 JAPAN

Keywords: one-man pilot, layout in the cockpit

1. Abstract

Manned research submersible *SHINKAI 6500* was completed in Nov. 1989, and has been operated by Japan Marine-Earth Science and Technology (JAMSTEC). She has capability to 1472 dives around the world deep sea. In 2012, JAMSTEC completed the first major upgrade of her, which was the largest scale since a launch.

In this annual overhaul, JAMSTEC has initiated the second major upgrade of her. This report will introduce some topics of *SHINKAI 6500* update. The first, the inside layout of a pressure hull will be changed, because of the one-man pilot's operation test dives and science cruises for 2017. The second, *SHINKAI 6500* has an underwater navigation system. This system calls "Shinkai-track".

2. Background

SHINKAI 6500 has capability of the depth to 6500 m. The crew complement is 3 persons, requires a pilot and a co-pilot, and an observer. R/V *YOKOSUKA* is a support Vessel for *SHINKAI* 6500. *SHINAKI* 6500 and *YOKOSUKA* have traveled many oceans such as the Pacific ocean, the Atlantic ocean, and the Indian ocean.

Nowadays, the technique of the sea floor investigation diversifies, the investigation's dive to which the observers are increased is strong and desired. Therefore, the capacity of the observer was increased from the present one person and the thing set to two persons was considered.

— Fig.1 *SHINKAI* 6500

3. Function of the Improvement

3.1 Purpose

After a consideration whether this improvement will be realistic, however more considerations as below 2 points were necessary.

The first one is function improvement. The second one is change in the layout of inside the pressure hull.

3.2 Underwater navigation system

First, we have begun to wrestle about function improvement of *SHINKAI 6500.*

SHINKAI 6500 usually uses the SSBL (Super Short Base Line) system. But the acoustic position does not tell automatically from mother ship to *SHINKAI 6500.* When the pilot wanted to know their position, they had to ask to an acoustic navigator onboard R/V YOKOSUKA by underwater communication system. Therefore, the new track positioning indication system was developed and installed in a few years ago. This system uses the DVL only, the conventional acoustic positioning system is not required.

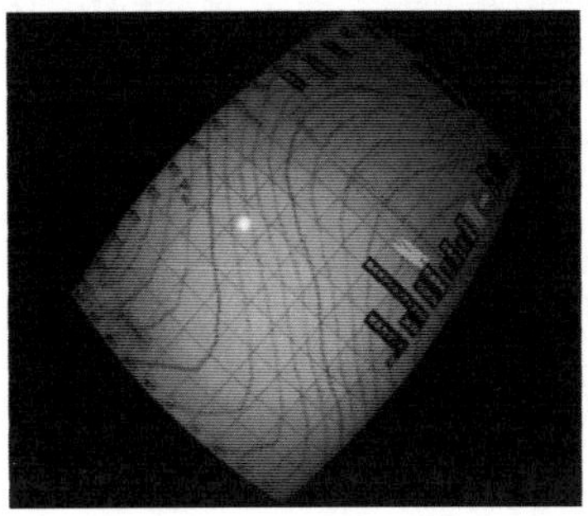

Fig.2 Underwater navigation system

Moreover, this underwater navigation system will be added an inertial navigation system (INS) to install in *SHINKAI* 6500 in this upgrading. This INS consists of a high accurate inertial sensor and Doppler sensor, and enables detecting with high accuracy the position, speed, azimuth, and attitude underwater. These positioning systems are using a tablet PC. Information such as a target direction and the distance until the target can be grasped by the thing which makes a tablet PC indicate the track and seafloor map.

However, the INS system usually has cumulative errors, and needs to update the position to clear this error. Therefore, we will consider to use the down link system by using the underwater communication system, and be shown the down link data on a tablet PC.

3.3 Cockpit

A titanium alloy pressure hull is equipped on *SHINKAI 6500*. The inner diameter is 2m, and thickness is about 74mm. *SHINKAI 6500* has 3 view ports, pilots uses the center view port for a maneuvering, and an observer uses the left side view port.

However, it is difficult to using the starboard side view port for an observation very much, because of an emergency oxygen gas cylinder and first-aid breathing system and the other equipment are installed. So, it is not suitable for long observation, and if the observer increase from 1 person to 2 persons as it is, the observation environment is not so good and the effect is not obtained. Therefore, the placement of instruments was arranged so that the copying machine sitting behind the pilot was easy to operate and monitor.

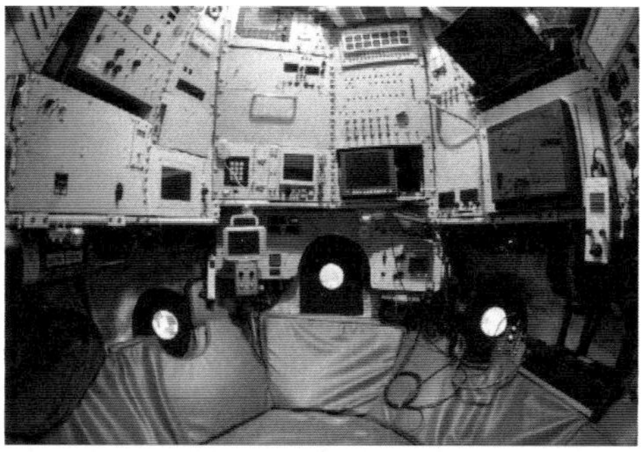

Fig.3 Inside of the pressure hull (present style)

3.4 Consideration of a layout

It is important that the starboard view port side arrangement will be improved and to be able to operate and monitor of instruments by a pilot. Then, it is necessary to fit for the space and lay out about a monitoring and operation carried by oneself up to now. In particular, relocate important equipment for monitoring to the vicinity of the pilot operating around the front view port.

From 2014, we started to make a new frame of inside the pressure hull and new arrangement for the one-man pilot operation was considered using mock-up.

Then, it is necessary to consider from each pilot opinions and suggestions. We spent a lot of time on the discussion step by step. After many trying and consideration, we decided a new layout of inside the pressure hull in 2015. From 2016, we started to carry the temporary cables and consideration of arrangement the instruments.

Fig.4 Mock-up of a frame in the cockpit

This picture is the front hemisphere and the rear hemisphere of mock-up the frame in the cockpit (Fig.4). By miniaturizing and integrating the equipment in the cockpit, we made the layout to fit most of the equipment in the northern hemisphere of the pressure hull.

3.5 Pre-dive on the mock-up

Next step is about the pre-dive check of the new style of the cockpit.

A pilot prepared various scenarios and operated about the phenomenon which occurred along a scenario, the improvement points was checked and advanced. As a result of the pre-dive check on the mock-up, a problem of 46 and a proposed measure of 98 were indicated.

The problem which had was to confirm how a pilot could put a double-check by a co-pilot could be put into effect with a pilot. Reconsideration of a check procedure, confirmation from the outside staff, mechanization and a proposed measure by support equipment were considered about these problems. As the result of these discussions and check to trying a fit for the space, we decided a new layout in the cockpit.

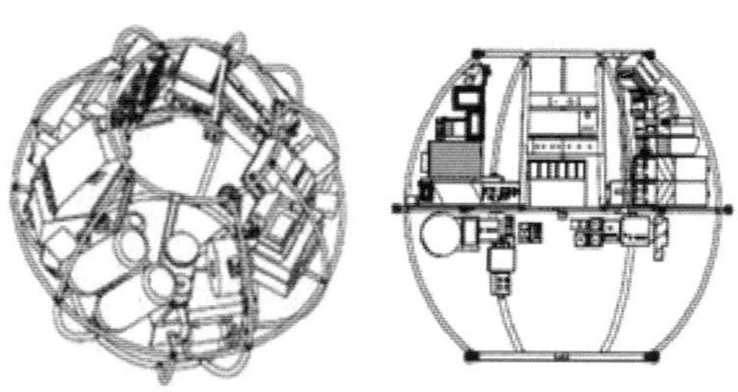

Fig.5 Mock-up of a layout image in the cockpit

3.6 Apply for the permission

In addition to explaining the operation of the one-man pilot to the Ministry of Land, Infrastructure,

Transport and Tourism concerned, we are continuing consultations to approve the change of the maneuvering manual and are proceeding with approval procedures. And also, it is necessary to get the certification of the Japanese ship classification society (ClassNK) for using the updated equipment. When the navigation permission is issued at the test dive after the completion of construction in March, the operation of the one-man pilot legally

Since it will be approved, we are continuing work including inspection work for this.

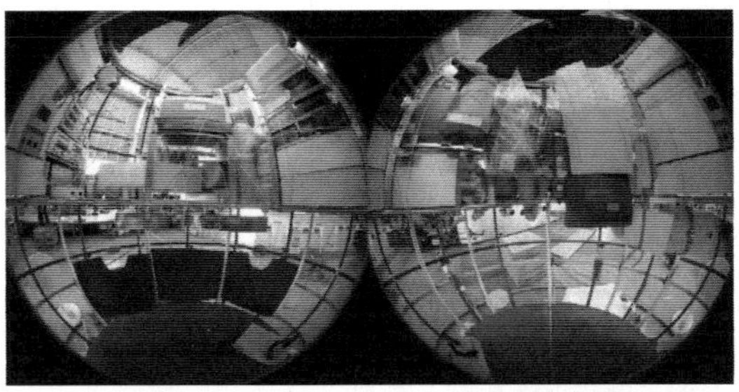

Fig.6 Mock-up of a layout in the cockpit

3.7 Consideration of operation restriction

Regarding the operation of the one-man pilot, depending on the contents of the dive survey, it is assumed that it may be more efficient to diving with two pilots.

The assumed dive is the following case. For example, a dive survey in hydrothermal vent fields, in the past where the rapids in the seafloor are observed, When the undulation of the seafloor is intense and it is assumed that it is necessary to strengthen monitoring.

On the other hand, the possible dives for one-man pilot diving are as follows. For example, dive survey that kept the height sufficient, survey area where flat seafloor or gentle slope is expected, mapping and geological survey.

In implementing the one-man pilot dive, it is assumed that the observer is requested to operate the equipment used for the survey observation limitedly. We will confirm and decide which equipment will be targeted by future training dives.

4. Conclusion

Remodeling of a layout in the cockpit has started in about 4 months' plan from November. After it was considered variously, it was possible to settle more problems concurrently with function improvement of equipment and decide about a new layout in the cockpit.

After the completion of this work, an experimental dive cruise is planned on March. Thereafter, a comprehensive training dive cruise for one-man pilot operations is planned on April. In addition, we will conduct actual training dives and extract problems and improvements for various situations for one-man pilot operation, and we are planning to confirm whether the equipment arrangement is appropriate and operation is not hindered.

After April, we plan to carry out necessary remodeling and adjustment from the information obtained by the training dive and carry out a survey dive with the conventional two pilots for one year. From 2018, we plan to start operation on survey diving that can be handled by one-man pilot operation.

Factors for Success of U.S. Research Submersible Operations

Underwater Intervention

Manned Undersea Vehicles

Dr. Karen M. Kohanowich

February 21, 2017

From Cousteau to Cameron: A Quadrant Model for Undersea Marine Research Infrastructure (UMRI) Assessment

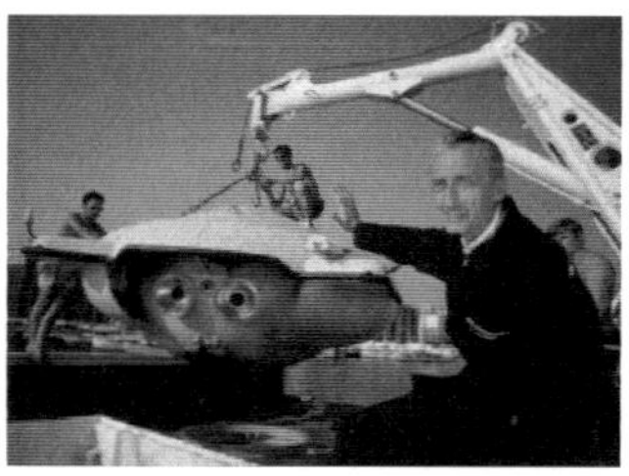

George Mason University dissertation

Karen M. Kohanowich

November 8, 2016

Available on ProQuest in Spring 2017

Using an Undersea Marine Research Infrastructure (UMRI) Quadrant Model to Examine U.S. Manned Research Submersible Operations

- Background
- The Model
- Factors
  - Technical
  - Operational
  - Functional
  - Societal
- Research Submersible Case Studies
- Recommendations

U.S. Submersible Science Infrastructure Programs

NOAA's
Undersea Research Program (NURP)

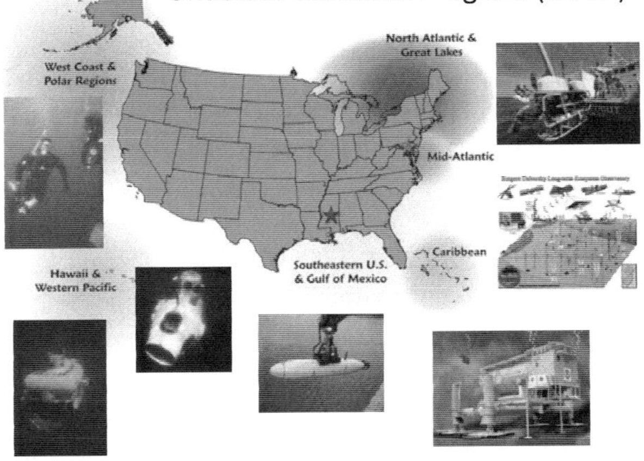

NSF's
National Deep
Submergence Facility
(NDSF)

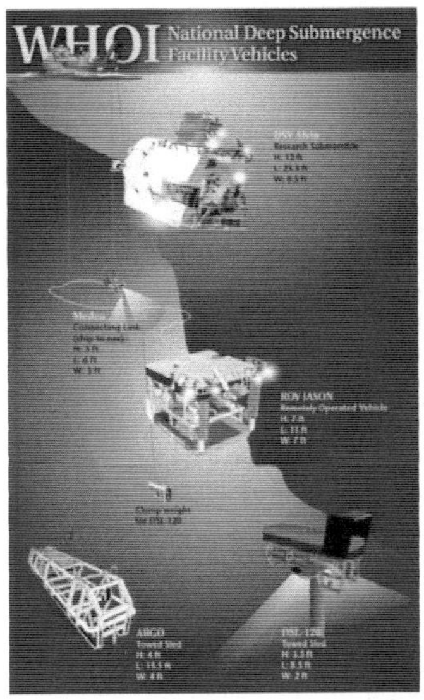

U.S. Submersible Science Infrastructure Reports

1956: Aspects of Deep Sea Research (Undersea Warfare)
1959: Oceanography 1960 to 1970 - Ch 7: Engineering
1965: Undersea Vehicles for Oceanography
1974: Future Facilities Requirements
1979: Research Submersible Requirements
1981: Assessment of Federal Technology for Ocean Research and Monitoring
1982: Submersible Science Study
1986: Marine Research and Low Cost ROVs
1990: Submersible Science Study
1994: The Global Abyss (DESSC)
1996: Undersea Vehicles and National Needs
1999: DESCEND (DESSC)
2004: Future Needs in Submersible Science
2011: Critical Infrastructure for Ocean Research and Societal Needs
2015: Sea Change: Decadal Survey of Ocean Sciences
2016: DESCEND 2 (DESSC)

National Academy of Sciences

Federal

UNOLS
University National Oceanographic Laboratory System

DESSC
DEep Submergence Science Committee

DESCEND
DEveloping Submergence Science for the Next Decade

"Science, Technology, and People"

Factors that might be important ingredients in the success of a science or technology project:

- ***People*** *(including tenacity/passion, curiosity, courage, vision), and things related to people*
- ***Money***
- ***Serendipity*** *(favors the trained, open mind)*
- ***External Factors*** *(politics, opportunities/circumstances, other science and technology developments; things you do not control)*

OceanGeeks

From: 2003 Briscoe, *Oceanography Vol. 16 Nr. 1*

Undersea Marine Research Infrastructure (UMRI)
Quadrant Assessment Model

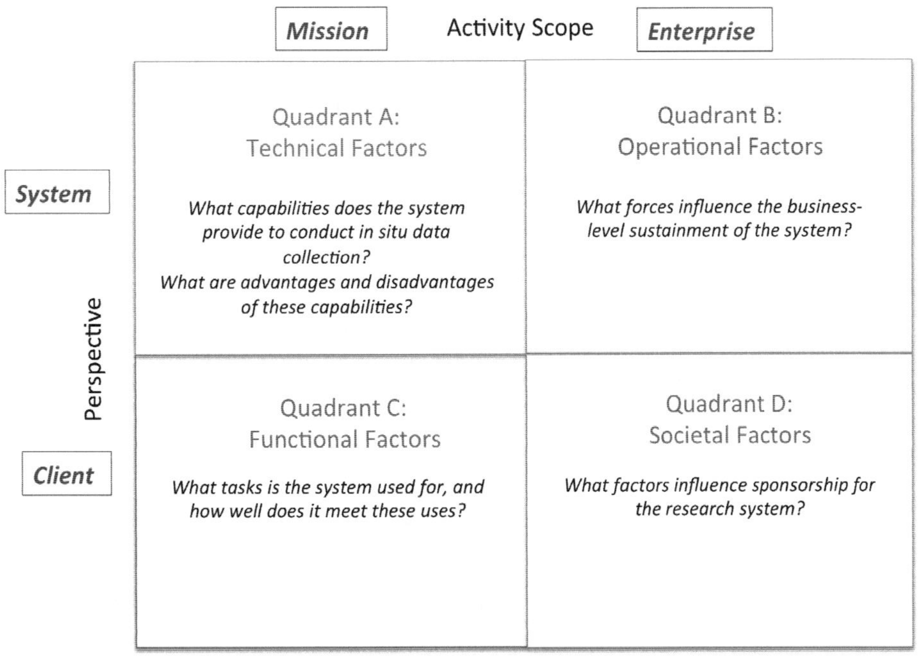

Model Illustration

Attributes of Factors

DEGREE: What is the magnitude of the impact on sustainment?

CHARACTER: Is the impact positive or negative?

RELATIONSHIP: Is the impact of the factor related to other factors?

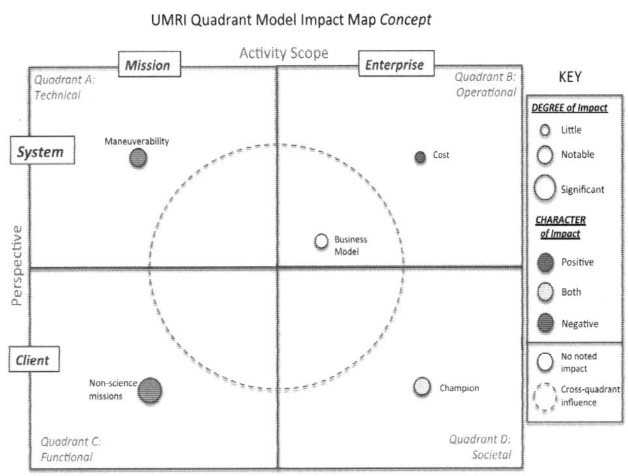

Data and Resources

- Literature
  - Journals, books, magazines, conference papers, websites
- Submersible logs
- Engagement (>60 stakeholders)
  - Targeted interviews (29)
  - Opportunistic conversations
  - Email and follow up
- Range of experience
  - Scientists, managers, operators
  - HOV and ROV experience
- *What has replaced the HOV?*

MTS marine technology society

Quadrant A: Technical Factors

System, Mission

- Maneuverability
- Time on Site
- Connectivity
- Structure
  - Ruggedness
  - Buoyancy Control
  - Noise
- Vision
- Being There

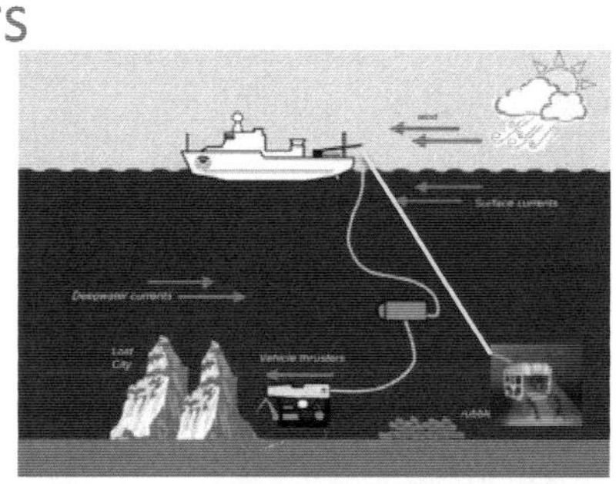

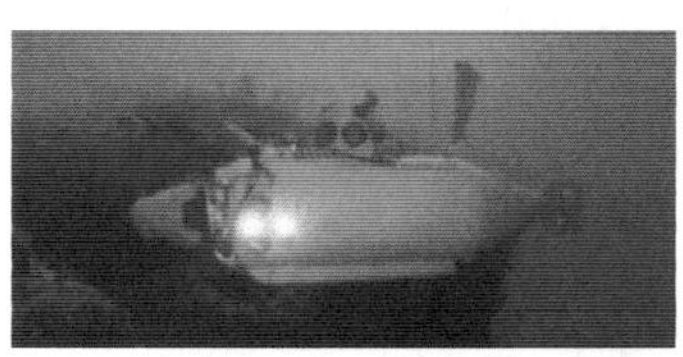

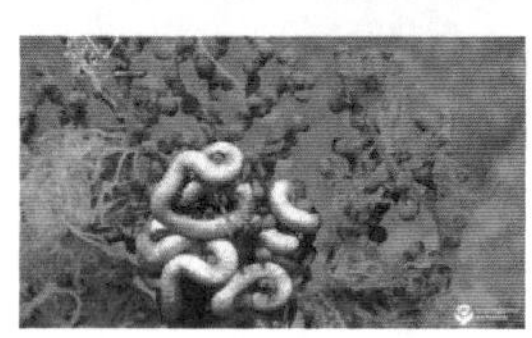

"Being There"

"Real Time Intellectual Resolution"

- React
- Explore
- Field work
- Physical discomfort
- Immersion in the task

"The particular skill of humans in noticing anomalous or emergent features and events and rapidly scanning an environment for sought features is what continues to give humans an edge over robots in the context of exploratory science ..."

(2014, National Academy of Sciences Space Studies Board)

Quadrant B: Operational Factors

System, Enterprise

- Cost
  - Capital
  - Operating
- Operating Team
- Safety
  - Certification
  - Record
- Availability
  - Geographic
  - Reliability
- Business Model

Quadrant C: Functional Factors

Client, Mission

- Activities
  - Micro ecosystem assessment
  - Macro ecosystem assessment
  - Transect
  - Mid-water studies
  - Other missions
- Continuity
- Personal Preference
- Environment

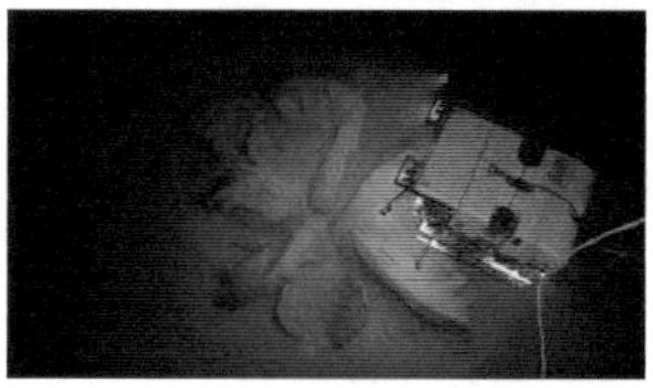

Quadrant D: Societal Factors

Client, Enterprise

- Sponsors
  - Government
  - UNOLS
  - Academic
  - Private
  - Champion
- Science Priorities
- Inspirational
- Risk Perception

Inspirational

- National Pride
- Recruitment

"although the effect on students can be used as a rationale for some spending on ... the human spaceflight program, few in the education realm would view this effect itself as a rationale for funding the human spaceflight program" (National Academy of Sciences Space Studies Board, 2014, p.61)

"If you'd have told me when I was a kid that 'oh, someday you'll grow up and you can sit in front of a computer screen and watch', I don't know that I would have been inspired to go. I was watching Jacques Cousteau out on the ship and diving and stuff. But things are different now and I don't know if it has the same impression" (Scientist M).

UMRI Quadrant Model Impact Map *Template*

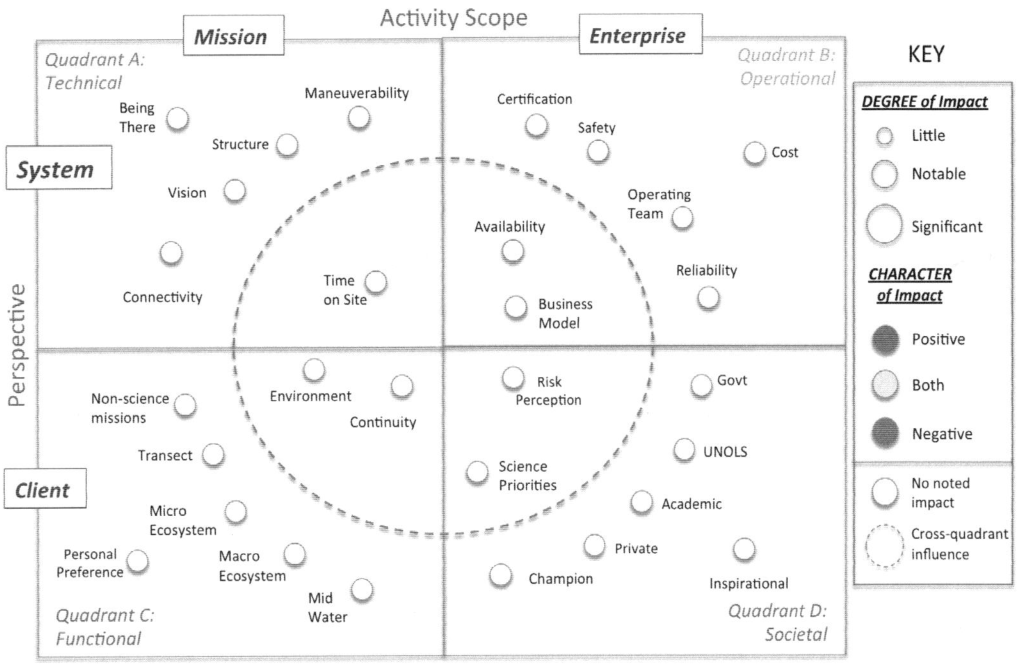

ALVIN
4500m

- NSF Funded
- UNOLS/DESSC

Woods Hole Oceanographic Institution (WHOI)

PISCES IV/V
2000m

- NOAA Funded
- Non-UNOLS

Hawaii Undersea Research Laboratory (HURL)

DELTA
100m

- Private industry
- Strong *ad hoc* user community

DELTA Oceanographics

JOHNSON SEA LINK I/II
1000m

- Private endowment
- Grant funded
- State university

Harbor Branch Oceanographic Institution at Florida Atlantic University (HBOI) (FAU)

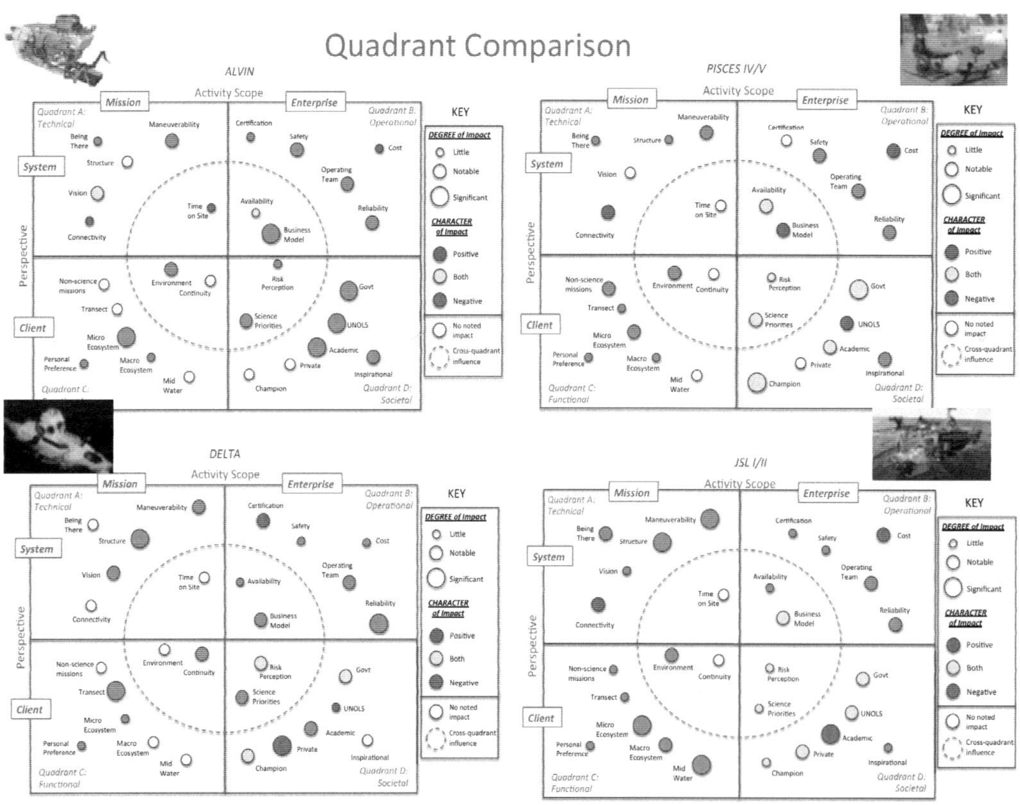

Case Study Highlights

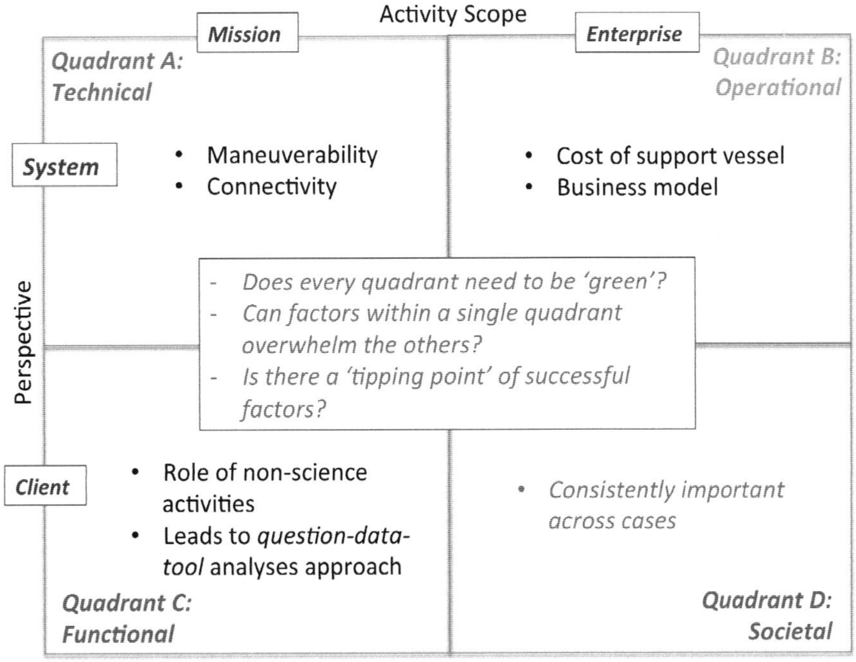

The Model as a Framework for Future Study

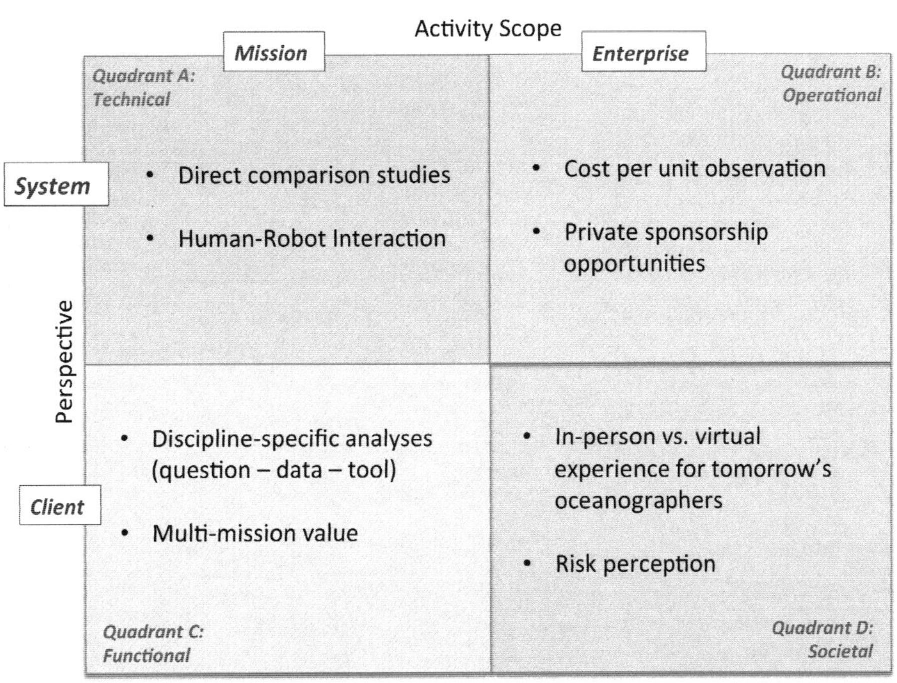

Translation to Action

UMRI Model Actors

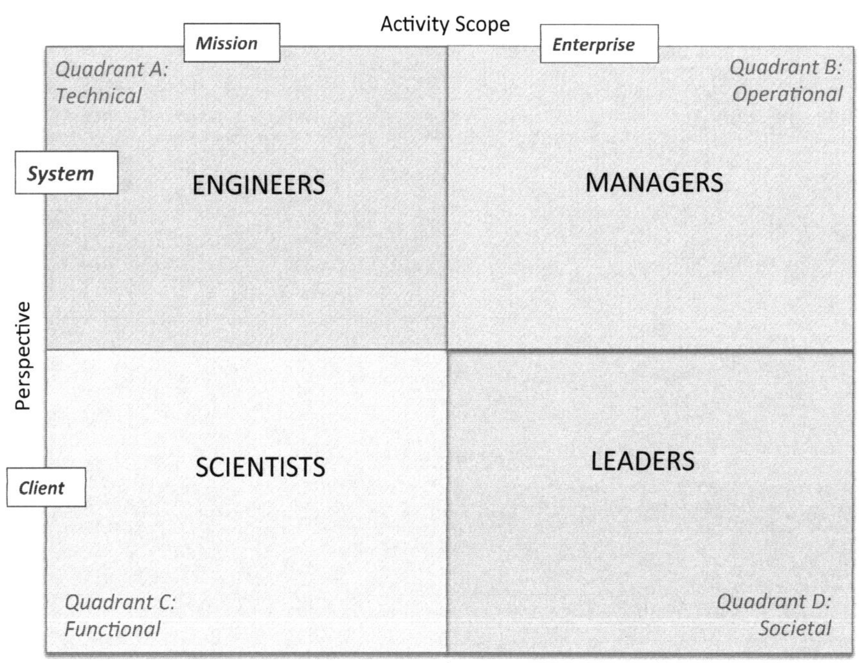

High Pressure Sealing

Lutz Mueller

Glenair.

High Pressure Sealing

Multiple Techniques

- Glass-to-Metal
- Welding/Brazing
- Potting/Adhesives/Sealants
- *Elastomeric Mechanical*
  - *Boot Seals*
  - *O-rings*

Glenair.

Glass to Metal

Robust Sealing and Isolation

- Hot glue Gun – except with glass
- Compression Seal (Pressure Seal)
  - CTE mis-match
  - Chemical and Mechanical bond
- Matched Seal (Hermetic Seal)
  - Matched CTE
  - Chemical bond

Matched Seal

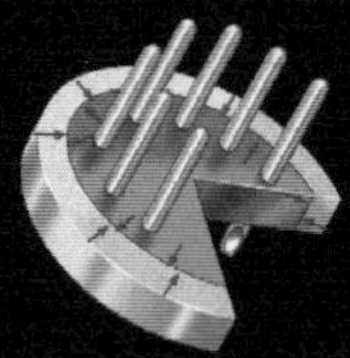

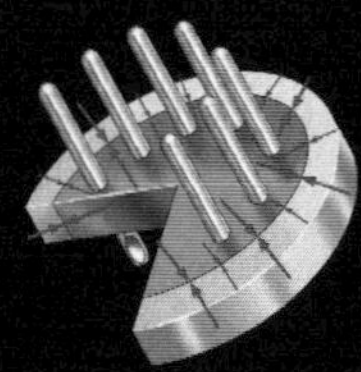

Compression Seal

Glenair.

Glass to Metal

Robust Sealing and Isolation

- Process sensitive
- Quality Control
  - Helium Leak Check
  - Pressure Test
  - Visual Que's
- Strong and Robust
- Cons:
  - Expensive (tooling/fixturing, multiple process)
  - Material Restrictions (due to heat and chemistry)

Glenair.

Welding/Brazing

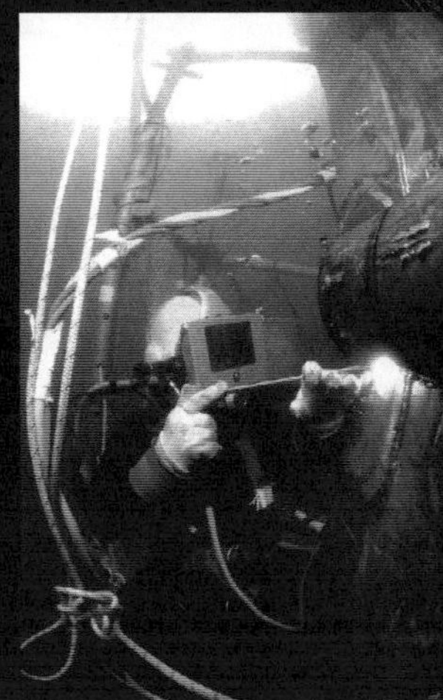

- Brazing
  - Does not melt base material
  - Types (Torch, Dip)
- Welding
- Joining to parts by melting of base material
- Welding types – Generally form factor dictates
  - Arc ("Stick",TIG, MIG)
  - Gas (Oxy Fuel)
  - Resistance (Spot)
  - Beam (Electron, Laser)
  - Solid State (Friction, Explosion, Diffusion bonding)

Glenair.

Welding/Brazing

- Process sensitive
- Quality Control
  - Helium Leak Check
  - Pressure Test
  - Visual Que's
  - Quality Assurance with certified welder
- Very Strong, Very Robust
- Cons: Heat transfer, material restrictions, expensive

Glenair.

Potting/Adhesives/Sealants

- Mechanical Seal blocking liquids
- Semi-Adhesive (chemical bond)
- Operating Temperature is critical
- Material Choice is critical
- Least Reliable:
  - False security
  - Structural is never flexible Flexible

Glenair.

Potting/Adhesives/ Sealant

- Process sensitive
- Quality Control
  - Thermal Cycling – Delamination due to CTE mis-match
  - Helium Leak Check
  - Pressure Test
- Fragile seal with respect to age and heat
- Cheap to implement
- Cons:
  - Unreliable

Glenair.

Elastomeric Mechanical Seals

Unsung Hero of the Sealing World

- Small, cheap, forgettable → CRITICAL
- Multiple use
- Performance dependent on design not process
- Conventional techniques (easy)
  - Manufacturing
  - Inspection
- Easily serviced
- Very effective when designed properly

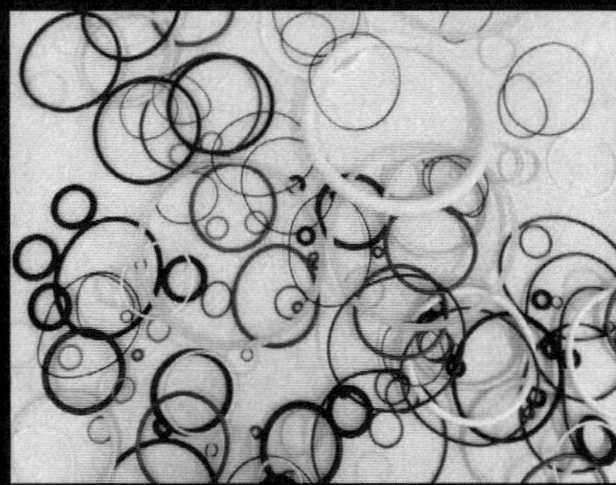

Glenair.

Proper Design

Gland Design

- Axial seal (face seal)
- Radial seal (piston or rod)
- Squeeze
- Extrusion gap
- Groove width

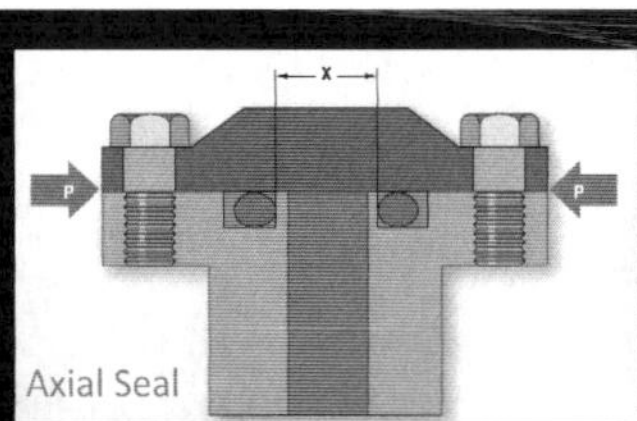

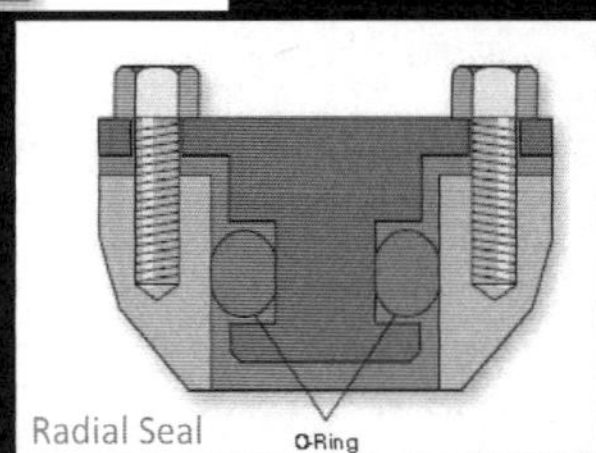

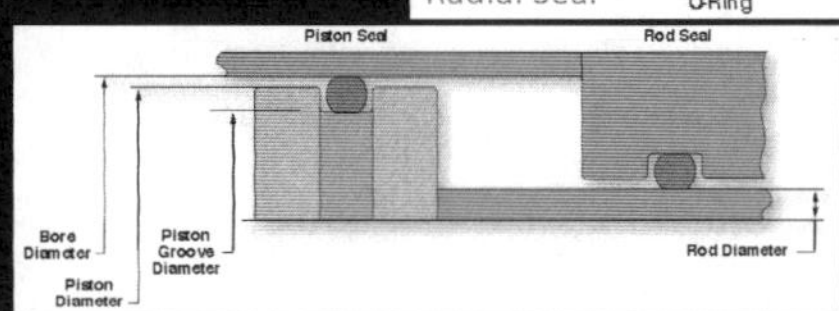

Glenair.

Proper Design

Material Choice

- Material impacts
  - Temperature
  - Fluid compatibility
  - Specialty (abrasion, gas decompression, conductive)
- Sub-Sea materials (-20°C- +80°C)
  - EPDM
  - Nitrile (HNBR or NBR)
- Hi-Temp – Fluorinated
  - Viton
  - Aflas

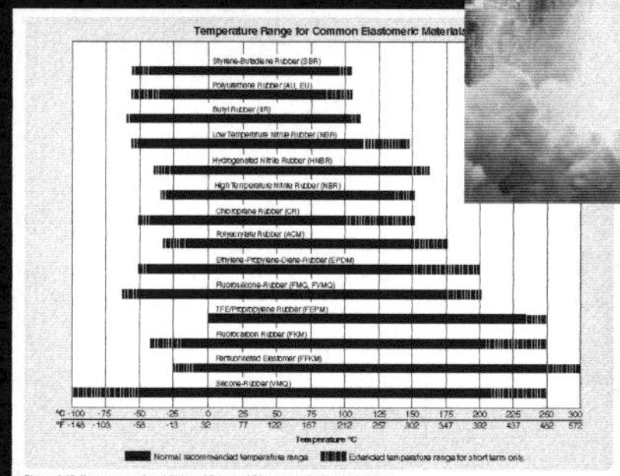

Temperature Range for Common Elastomeric Materials

Figure 2-25: Temperature Capabilities of Principal Elastomers Employed in Seals

Glenair.

Proper Design

Shore Hardness

- Soft shore advantages
  - Easier to assembly
  - Less friction forces
- Hard shore advantages - Prevents extrusion
- Extrusion is the KEY to high pressure sealing

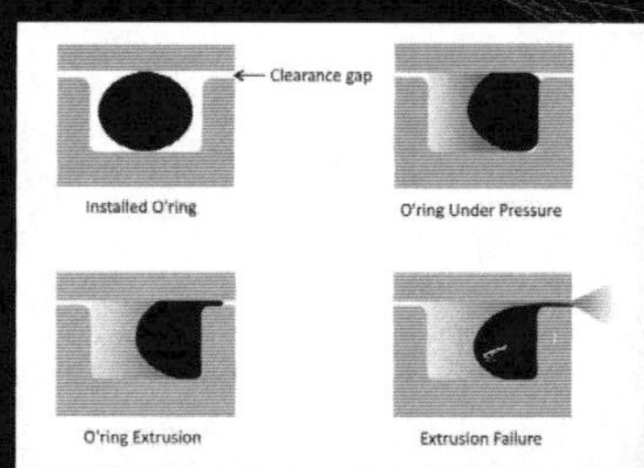

Glenair.

U. S. COAST GUARD
Glenair
High Pressure Sealing
Lutz Mueller

Making deep sea research affordable

Pisces VI

Self introduction of speakers
*Scott Waters
*Vance Bradley
*Grace C. Young

Scott- Specs of Pisces VI
*Originally built by HYCO in 1976 and the hull was produced in Austria
*Owned and operated buy IUC until warehoused in 1990
*3 occupants capacity
*2000 m operational / 2500m test
*16,000 lb weight
*47 KW power on board
*8 hour mission time with 120 hours emergency life support

Scott- Main re-design points were all focused on safety, cost, multipurpose abilities

Vance – Safety
*120 hours life support
*4500psi emergency ballast air
*350lb drop weight
*9khz UQC, 27khz UQC, VHF, Underwater Tracking system up to 5000m
*Jettisonable thrusters
*Strobe
*Emergency segmented power

Scott – Cost
*Acquisition cost
*Fits in two 20' Hi Cube shipping containers (Cost of shipping)
*Home base will operate from shore option to lower the cost of operation
*Only weighs 8 tons so cost of support ship is lower than bigger submarine
*Storage costs is low
*No large infrastructure employees are private contractors
*Shop built and fully owned then will be flipped for a profit (learned from home improvement industry)
*Many other cost savings techniques from my days running retail businesses both in the build phase and operational phase

Grace – Multipurpose abilities
*Six 4K cameras with digital recording ability and virtual reality ability
*HYCO Manipulator with basket (ability to add second arm)
*HYCO arm has multiple quick change tooling (multiple claws, water jet, and slurp gun)
*Six LED array lights ranging from 10,000 to 30,000 lumen
*Comes standard with laser scalier
*Optional geological push core sampler system, water sampling system, as well as small on call staff for specialized tooling

Vance – Design changes from the original Pisces VI
*Ballast system
*Electronics
*Thruster configuration (from 2 vectorable to 4 fixed)
*Battery system
*Trim weight / drop weight system
*Sail

Questions and Answers

Pilot Training Program of *Huan Dao Jiao Long* Submersible

Shuai Liu[1,2], Cong Ye[1,2], Lei Jiang[1,2], Liping Yu[1,2]
[1]China Ship Scientific Research Center (CSSRC), P.O. Box 116, Wuxi, Jiangsu,214082, China
[2]State Key Laboratory of Deep-Sea Manned Vehicles

Abstract: There are no certification measures of submersible pilot in China, before *Huan Dao Jiao Long* program. Now the trial certification measures for submersible pilot were published by Hainan Maritime Safety Administration of P.R.C, because of promotion of China Ship Scientific Research Center and the support of Maritime Safety Administration of the People's Republic of China. This paper initially review the process of the administrative permission for *Huan Dao Jiao Long* submersible, and then briefly introduce the trial certification measures for submersible pilot. Finally, theory training structure and practical training structure of the submersible's pilot are emphatically illustrated.
Keywords: Submersible, *Huan Dao Jiao Long*, Pilot, Training.

1Introduction

As a coastal country, China has jurisdiction over 3 million square kilometers of sea, continental coastline extends for about 18,000 kilometers, covering 11coastal provinces, autonomous regions and 53 coastal cities, with a huge development potentiality[1].

There is a good foundation of submersible manufacturing in China, in the past 20 years, more than 10 submersibles were constructed[2-3]. Although in 1987 & 1996 China Classification Society promulgated the "Rules for Building and Classing of Diving system and submersible[4]" (Submersible Rules).

There were no regulations on training assessment and certification of submersible pilot in China, so all submersible didn't allow to operate for commercial purpose before *Huan Dao Jiao Long* program. Now the trial certification measures for submersible pilot were published by Hainan Maritime Safety Administration of P.R.C, because of promotion of China Ship Scientific Research Center and the support of Maritime Safety Administration of the People's Republic of China.

2 Process of the Administrative Permission

China Classification Society (CCS) provides submersibles classification services which include the reviewing of design plans and surveys during construction. This is to verify the submersible project complies with CCS' Rules for Classification and other relevant standards.

Maritime Safety Administration of the People's Republic of China (China MSA) is the competent authority to exercise the administration of shipping safety and maintain the national sovereignty, including drafting and implementing national polices, laws and regulations and standards regarding waterborne traffic safety supervision, administrating the survey of ship, seafarers and pilots' training ,examination and certification and others[5].

Huan Dao Jiao Long submersible is designed and constructed by China Ship Scientific Research Center (CSSRC). Hainan Yalong Bay Underwater World Company is owner of the submersibles. The development was started from 2012. When we started the *Huan Dao Jiao Long* submersible project in 2012 years, we have difficulties in old Submersible Rules and missing regulations regarding submersible implemented by Maritime Safety Administration.

In order to overcome these difficulties, we have done much work. The main process including:

1. CSSRC cooperated with CCS Jiangsu Branch, drafted "Rules and regulations for the construction and classification of *Huan Dao Jiao Long* submersible" (*Huan Dao Jiao Long* Rules), in order to overcome obsolete Submersible Rules difficulty.
2. CSSRC applied for pilot program to Jiangsu Maritime Safety Administration of P.R.C (Jiangsu MSA), then Jiangsu MSA reported to China MSA. According to the requirements of China MSA, we have done much work about safety analysis of the submersible, China MSA agreed with the pilot program in the end.
3. CCS submitted *Huan Dao Jiao Long* Rules to China MSA, *Huan Dao Jiao Long* Rules and pilot program have been examined and approved by China MSA in August, 2013. Hainan Maritime Safety Administration of P.R.C (Hainan MSA) is responsible for administrating the survey of the submersible, submersible pilot training, examination and certification and others.
4. Hainan MSA drafted the trial measures for submersible pilot in 2016, CSSRC drafted pilot training program of *Huan Dao Jiao Long* submersible, China MSA have approved these documents.

3 The Trial Certification Measures for Submersible Pilot

The trial certification measures for submersible pilot were published by Hainan MSA[6] in December 2016, so this is only applicable to the area under Hainan MSA. The trial certification measure requires that the program of pilot training and examination is formulated independently based on feature of every submersible. When the training program is approved by Hainan MSA, manufacturer or submersible operator will be responsible for training according to the program, under the supervision of Hainan MSA.

The requirements of the tourist submersible in the trial measure are as follows:

1. The gross registered tonnage of the tourist submersible is less than 100;
2. The power of the tourist submersible is less than 220 kilowatt.

In order to qualify for this training, applicants must each submit the following information for evaluation:

1. All applicants must be at least 18 and less than 60 years of age.
2. All applicants must meet the health standards of seafarers.
3. All applicants must have valid master certificate and have good safety records , actually served as the captain for 24 months.
4. Evidence of completion of the submersible pilot training.

When all training contents are completed, Hainan MSA organize examination. Exams including theory and practical operation, the score on theory examination must exceed sixty percent, full credit is one hundred percent. Only when applicants pass the theory and practical operation can applicants can get the submersible pilot certificate.

4 Training Program of the *Huan Dao Jiao Long* Submersible Pilot

The theory training is comprised of submersible theory and submersible management.

The theory training courses of submersible are 60 hours in total. The main class works include: overall arrangement, overall performance, internal outfitting, emergence procedures, pressure structure and seal, materials and anticorrosion, non-pressure structure, electrical power &

distribution, ballast & trim system, propulsion & maneuvering system, communications, navigation & monitoring, life support system and air conditioning system.
The theory training courses of submersible management include the following items: administrative statutes, operation limitations, pre dive and post check, pilot responsibilities and passenger management.
The practical operation training is comprised of submersible operation and submersible maintenance.
The submersible operation training include inspection and operation before lifting, surface operation & manoeuvres, diving operation and underwater maneuvering, emergency operation.
The submersible maintenance training include daily inspection, battery charging, compression air resupplying, compression oxygen resupplying, carbon dioxide absorbent replacing, periodic inspection of submersible.
The first batch of pilot of *Huan Dao Jiao Long* submersible took part in the theory and practical examination hold by Hainan MSA from January 19th, 2017 to January 20th, 2017.

References

1. Yu Liping, Ye Cong, Hu Zhen, Liu Shuai. Status and Prospects of Tourist Submersible Industry in China[C]//International Conference on Subsea Technologies. St-Petersburg, Russia: 2014: 239-246.
2. LIU Shuai, YE Cong, HU Zhen, ZHANG Xiaoxian; Status and Prospects of Tourist Submersibles [J]. Shipbuilding of China, 2013, 54 (03): 200-208. (in Chinese)
3. Zhang Guoguang, Dong Jianshun.The Current Situation of Industry and Its Safety Problem for Tourist Submarines in China[C]//Asia-Pacific Diving & Underwater Technology Conference. Beijing: Ocean Press, 2008: 274-279. (in Chinese)
4. China Classification Society. Rules for Building and Classing of Diving system and submersible[S]. Beijing: China Communications Press; 1996. (in Chinese)
5. Maritime Safety Administration of the People's Republic of China. Introduction [DB/OL]. [2017-01-09]. http://en.msa.gov.cn/index.php?m=content&c=index&a=lists&catid=328.
6. Hainan Maritime Safety Administration of P.R.C. Normative Document [DB/OL]. [2017-01-09]. http://hnmsa.gov.cn/u/cms/www/201612/0817043816go.pdf.

The 14th Annual MTS MUV, UI 2017

中船重工

Research & Application of One Man Atmospheric System

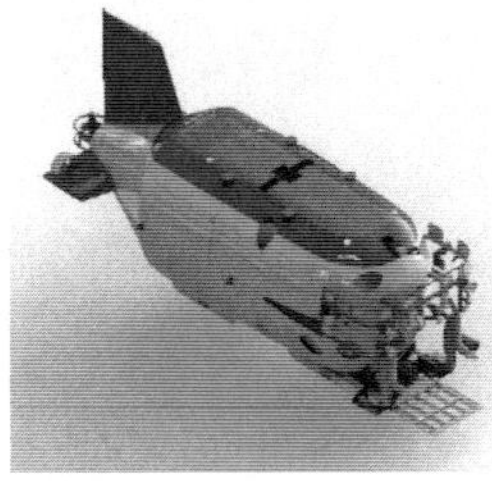

China Ship Scientific Research Center

2017.1

DIRECTORY

中船重工

1、OVERVIEW

2、SYSTEM CONFIGURATION

3、MAIN SPECIFICATIONS

4、RESEARCH & MANUFACTURE OF OMAS

5、EXPERIMENTAL VERIFICATION

6、SUMMARY

Manned Submarine

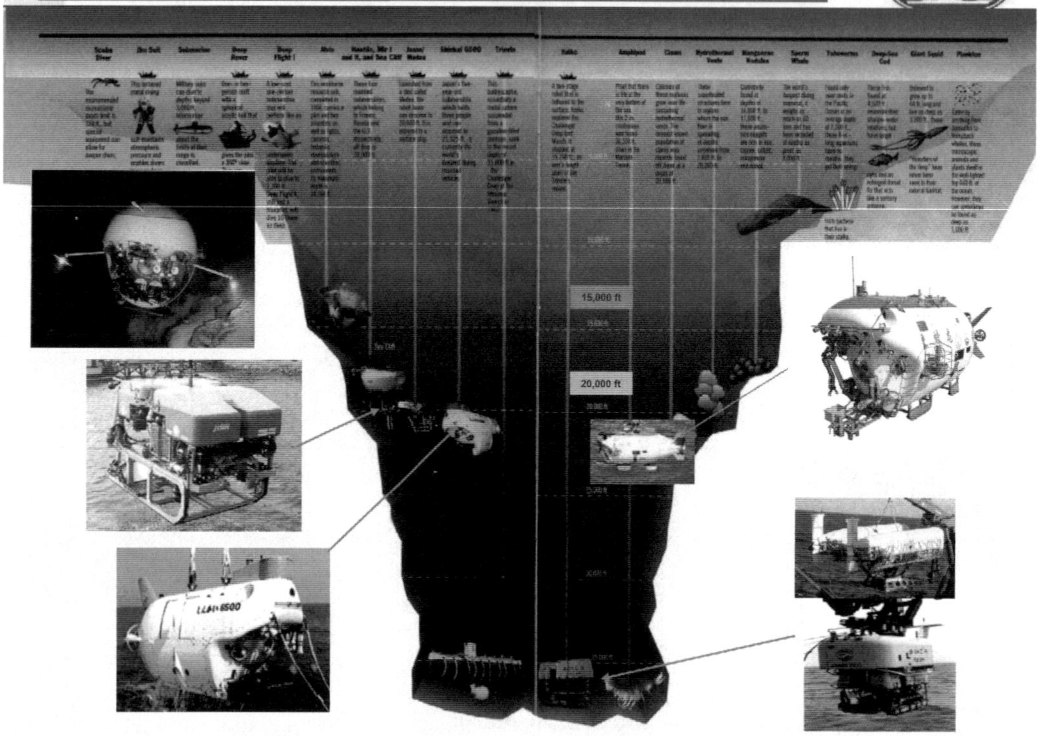

1. OVERVIEW

中船重工

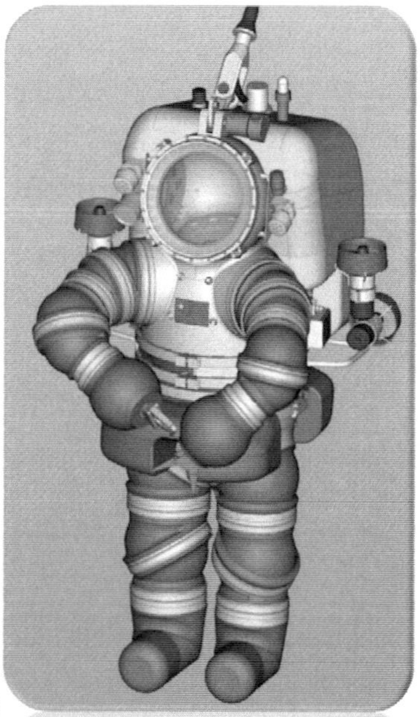

One Man Atmospheric System (OMAS), is an articulated anthropomorphic single person submersible which consists of the Suit, Tether Management Systems (TMS), Launch and Recovery Systems (LARS) and Surface Control Systems (SCS). The system has the capability of allowing a pilot operating at the working site underwater. OMAS maintains the internal pressure at one atmosphere. There is no need for decompression schedules and it increases the work times. Meanwhile, the pilot is put directly at the working site where they can observe directly and control tools dexterously. So, OMAS has more effective than other traditional underwater vehicles. The images in operation which are transferred synchronously to the surface can be used to monitor and guide the operation.

1. OVERVIEW

中船重工

OMAS has a long development history. At present, America, Canada, Russia, Italy have already developed related products.

HardSuit

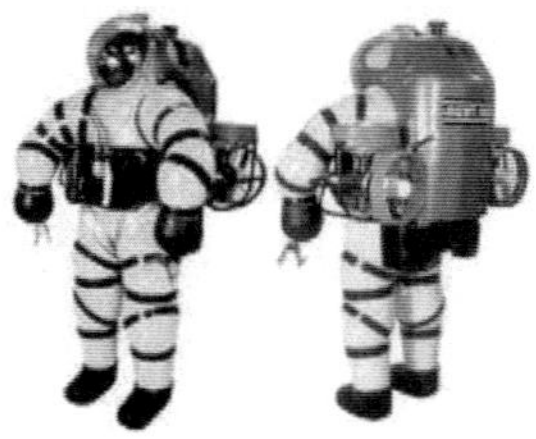

NEWSUIT

HardSuit2000

1. OVERVIEW

中船重工

Table of OMAS around the world

	America	France	Italy	Russia
Model	HS2000	HS1000	HS1200	HS1200
Working Depth	600m	300m	365m	365m
Height (m)	2.08	2.06	2.06	2.06
Weight (kg)	436	372	378	378
Thrust Power	2×2.25HP	2×2.25HP	2×2.25HP	2×2.25HP
Housing Material	Wrought Aluminium	Cast Aluminium	Cast Aluminium	Cast Aluminium
Life Support	8 h (normal), 48 h (emergency)	8 h (normal), 48 h (emergency)	8 h (normal), 48 h (emergency)	8 h (normal), 48 h (emergency)

1. OVERVIEW

CSIC
中船重工

OMAS is an important equipment for submarine underwater rescue. In China, OMAS is also an important equipment for the deepsea emergency maintenance .

This OMAS is developed to meet the needs of maritime emergency maintenance in 2010, and it's deepest diving depth is 502m at sea trail in 2015.

DIRECTORY

CSIC
中船重工

2. SYSTEM CONFIGURATION

中船重工

OMAS consists of Suit, Neutral Umbilical Tether, Tether Management Systems (TMS), Armoured Cable, Launch and Recovery Systems (LARS) and Surface Control Systems (SCS).

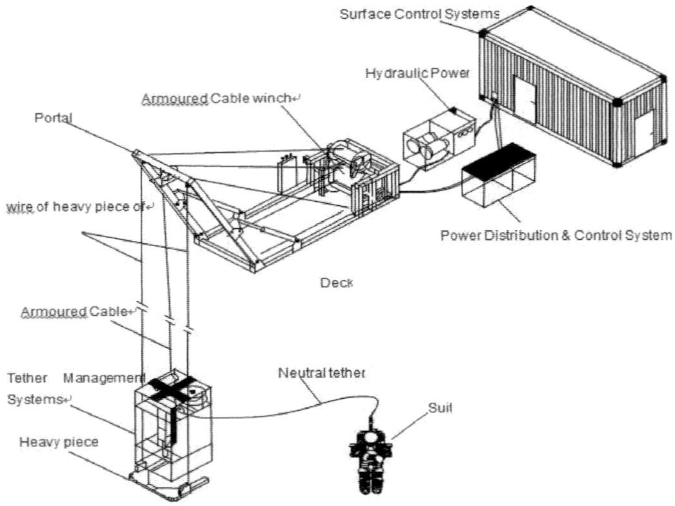

DIRECTORY

中船重工

1、OVERVIEW

2、SYSTEM CONFIGURATION

3、MAIN SPECIFICATIONS

4、RESEARCH & MANUFACTURE OF OMAS

5、EXPERIMENTAL VERIFICATION

6、SUMMARY

3. MAIN SPECIFICATIONSN

中船重工

OMAS's main specifications:

a) Maximum working depth: 500 meters;
b) Working Conditions: recycling ≯ level 4, operation ≯ level 3;
c) Main Dimension of Suit: overall height about 2.1 meters, overall width about 1.3 meters ;
d) Suit weight: less than 550 kg in air (not including the driver), about 2-4 kg in water (including driver);
e) Life Support: 6 hours (normal), 48 hours (emergency);
f) Maximum forward speed ≥1.0 knot, maximum up/descend speed ≥1.5 knot;
g) Communications: wired intercom, acoustic underwater phone

DIRECTORY

中船重工

1、OVERVIEW

2、SYSTEM CONFIGURATION

3、MAIN SPECIFICATIONS

4、RESEARCH & MANUFACTURE OF OMAS

5、EXPERIMENTAL VERIFICATION

6、SUMMARY

4. RESEARCH & MANUFACTURE OF OMAS

中船重工

- **2010 Start**
- **2015 done**
- **2015 Sea Trail**

Maximum working depth: 500m

Deepest diving at sea trail: 502m

4. RESEARCH & MANUFACTURE OF OMAS

中船重工

Body consists of helmet, upper body, waist adapters, lower body and limbs.

Helmet

The shape of helmet's glass section is like spherical cap with a 150 degree field of view. Helmet is connected with upper body by bolts, constituting pilot's head space.

Upper Body

Upper body has four openings. The top opening is connected with helmet.

Lower Body

Lower body has three openings. The top opening is connected with waist adapters. And the two bottom symmetrical openings are connected with right and left legs.

Waist Adapters

The standard waist adapters have two specifications. One has the height of 50mm while the other one has the height of 100mmheight adjustment.

Limbs

Limb segments are connected by hydraulic pressure compensating rotary joints. Each arm assembly contains five rotary joints and each leg assembly contains three rotary joints.

Body

4. RESEARCH & MANUFACTURE OF OMAS

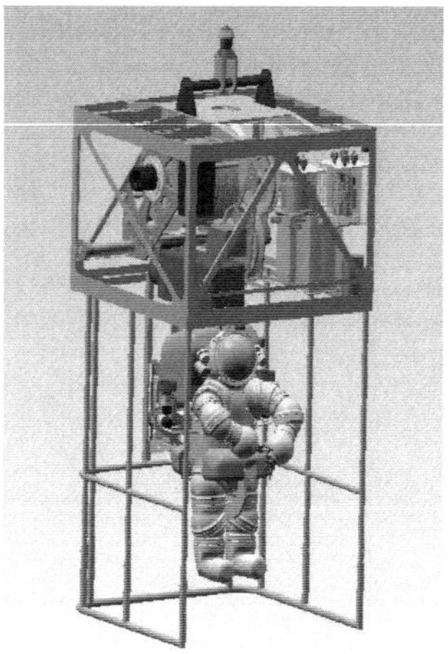

Tether Management Systems (TMS)

4. RESEARCH & MANUFACTURE OF OMAS

Launch and Recovery Systems (LARS)

4. RESEARCH & MANUFACTURE OF OMAS

Outside of SCS | Inside of SCS

Surface Control Systems (SCS)

4. RESEARCH & MANUFACTURE OF OMAS

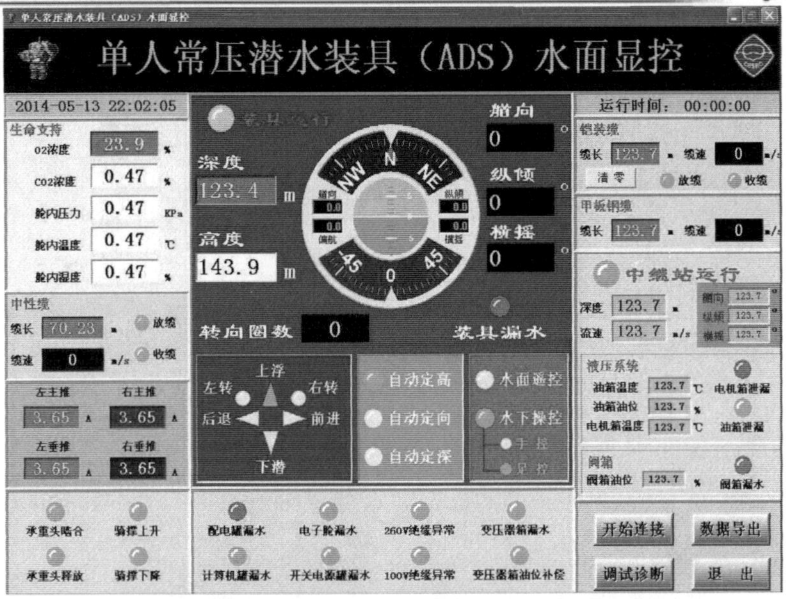

Control System

DIRECTORY

1、OVERVIEW

2、SYSTEM CONFIGURATION

3、MAIN SPECIFICATIONS

4、RESEARCH & MANUFACTURE OF OMAS

5、EXPERIMENTAL VERIFICATION

6、SUMMARY

5. EXPERIMENTAL VERIFICATION

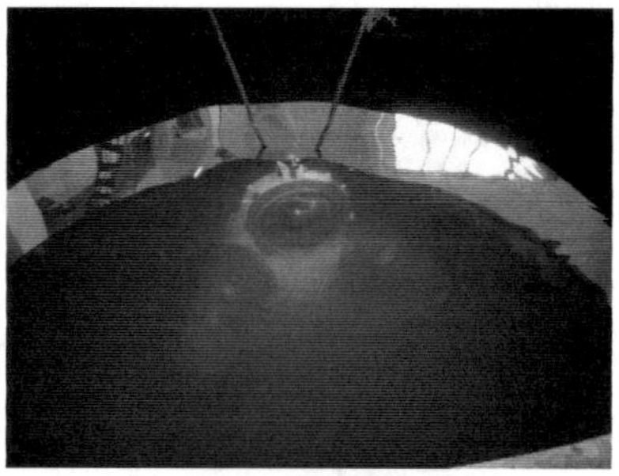

Pressure Test

5. EXPERIMENTAL VERIFICATION

中船重工

Life Support Test

5. EXPERIMENTAL VERIFICATION

中船重工

Basin Test

5. EXPERIMENTAL VERIFICATION

中船重工

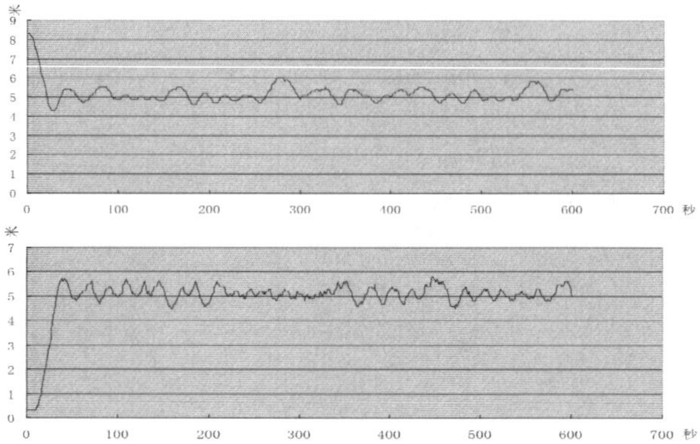

Auto Depth/ Auto Height

Sailing Test

5. EXPERIMENTAL VERIFICATION

中船重工

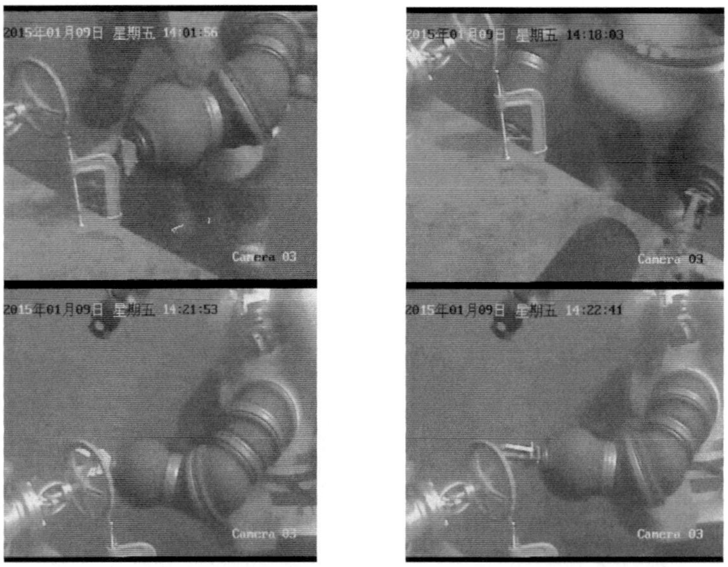

Water Tank Test

5. EXPERIMENTAL VERIFICATION

中船重工

Sea Trail

DIRECTORY

中船重工

6. SUMMARY

中船重工

In the offshore oil and gas industry, diving comes into widespread use in drilling, platforms building and wells exploring. The OMAS has been used in the high-altitude inspection and maintenance, underwater drilling installation, pipeline inspection and underwater exploration. And, due to its atmospheric pressure design, the OMAS can be used in cold and dangerous environment. Some of the proven application of the OMAS in the offshore oil exploitation are:

(1) Pipeline Installation and Detection.

(2) Underwater Objects Location and Recovery.

(3) Platform Observation and Waste Clearance.

(4) Scientific Exploration and Research.

The main powers' navies are equipped with excellent OMAS with routine access to the seafloor primarily in the 300-610m depth range. The OMAS are participated in the submarine rescue drill.

中船重工

THANKS !

Re-Classing Existing Submersibles

Jess G. Totten | 22 February 2017
Underwater Intervention Conference

ABS

Submersibles over 20 years old

- Will not be considered for Class
- No exceptions

ABS

Submersibles less than 20 years old, previously ABS classed

- Owner contacts their local ABS Survey Office
- Survey office contacts Divisional Survey Department
- Divisional Survey issues instructions for a Preliminary Condition Survey to consider the vessel for Re-Classification

ABS

Pre-condition Survey

- General internal and external examination
- Condition documented with photographs
- Service history and dive record review
- Review viewport documentation
- Additional examinations as needed or directed by Divisional Survey Office
- No charge to owner

ABS

Outcome of Pre-condition Survey

- Surveyor makes recommendation to Divisional and Corporate Survey Department whether or not to class the sub
- Engineering office may be asked for their input
- Corporate Survey Office, with input from attending surveyor and Divisional Survey Office, makes the final decision to proceed with Re-Classing

ABS

Next steps

- Divisional Survey Department will issue instructions for re-classing the Submersible
- Owner must work with local Survey office to complete all required Re-Classification survey requirements
- Any modifications must be submitted to the Technical office for review
  - Modifications performed while out of class
  - Modifications required to comply with the Rules
- Surveyor to witness any required repairs or tests, typically including a Special Periodical Survey based upon age
- Any due and overdue surveys as well as previous recommendations will need to be complied with

ABS

Special Periodical Survey

- Required for classed submersibles three years after class is granted, and at three year intervals thereafter
- Required of formerly ABS classed submersibles if the due date has passed
- Includes pressure testing of ballasting systems, life support piping systems and gas storage containers
- Viewports to be removed and inspected
- Gauges and monitoring instruments shown to have been calibrated recently
- Electrical systems examined
- Pressure hull inspected and NDT in highly stressed areas
- Test dive per requirements of Underwater Rules

ABS

Submersibles less than 20 years old, not previously ABS classed

- Preliminary Condition Survey.
- Process is similar to that for a previously ABS classed Submersible
- PVHO pressure boundaries and acrylic windows can only be accepted when built under survey by an IACS member Class Society or recognized Flag Administration
- Instructions from Corporate Survey Department will be based on the quality and completeness of documentation proving the submersible was built under survey
- Documentation includes, but is not limited to:
  - Welding procedures and performance qualification records
  - NDT records
  - Material mill test reports
- Special Periodical Survey is mandatory

ABS

What edition of the Rules applies?

- Agreed to between Owner, Corporate Survey Department, and Engineering office in each case
- For formerly ABS classed vessels, typically apply same edition of Rules in place when submersible was initially classed
- For non-ABS classed vessels, typically apply current edition of Rules

Owner responsibilities

- Complete a Request for Class (RFQ) form provided by local Survey Office
- Provide a signed affidavit detailing submersible's class status, modifications or repairs, and any other anomalies requiring Surveyor attention
- Provide all required documentation for the subject submersible, including class certificates, survey reports, viewport PVHO documentation, and manuals
- Submit details of any modifications for review by the Engineering office
- Complete all required repairs and modifications to the satisfaction of the attending Surveyor

Questions?

11 | Title of Presentation Goes Here

ABS

ABS

Thank You

www.eagle.org

Sea Trials of *Huan Dao Jiao Long* Submersible

Cong Ye[1,2], Shuai Liu[1,2], Dongsheng Wang[1,2], Deyong Hou[1,2], Lei Wang[1,2]
[1]China Ship Scientific Research Center (CSSRC), P.O. Box 116, Wuxi, Jiangsu,214082, China
[2]State Key Laboratory of Deep-Sea Manned Vehicles

Abstract: Acrylic-hulled submersible is able to maximize customer satisfaction of underwater viewing experiences, which is becoming the trend of the submersible. *Huan Dao Jiao Long* , China's first acrylic-hulled submersible, is designed and constructed by China Ship Scientific Research Center. The difficulties of obsolete domestic underwater vehicles rules and missing regulatory guidelines have been overcome. *Huan Dao Jiao Long's* development and sea trials were both accomplished in 2016. This paper initially reviews the course of the submersible, and emphatically illustrates items and contents of its sea trials. Finally, the follow-up development is prospected.
Keywords: Submersible, Acrylic-hull, Huan Dao Jiao Long, Sra trial.

1INTRODUCTION

Acrylic-hulled submersible is able to maximize customer satisfaction of underwater viewing experiences, which is becoming the trend of the submersible[1-3]. *Huan Dao Jiao Long* , China's first acrylic-hulled submersible, is designed and constructed by China Ship Scientific Research Center (CSSRC). Hainan Yalong Bay Underwater World Company is owner of the submersibles. The development was started from 2012 and now its design, construction, open water tank test and sea trials have been completed.

The purpose of this paper is to briefly introduce the process of development including design, construction, open water tank test and sea trial.

2 DESIGN CONSIDERATIONS

The submersible was designed to submersible classification of the China Classification Society[4] (CCS) with the pressure acrylic windows designed in accordance with American Society of Mechanical Engineers (ASME) Code for Pressure Vessels for Human Occupancy 1[5] (PVHO-1). The mission set for Huan Dao Jiao Long is that it can carry passengers with a sea depth of not exceeding 40 meters. The main design considerations are shown in Figure 1.

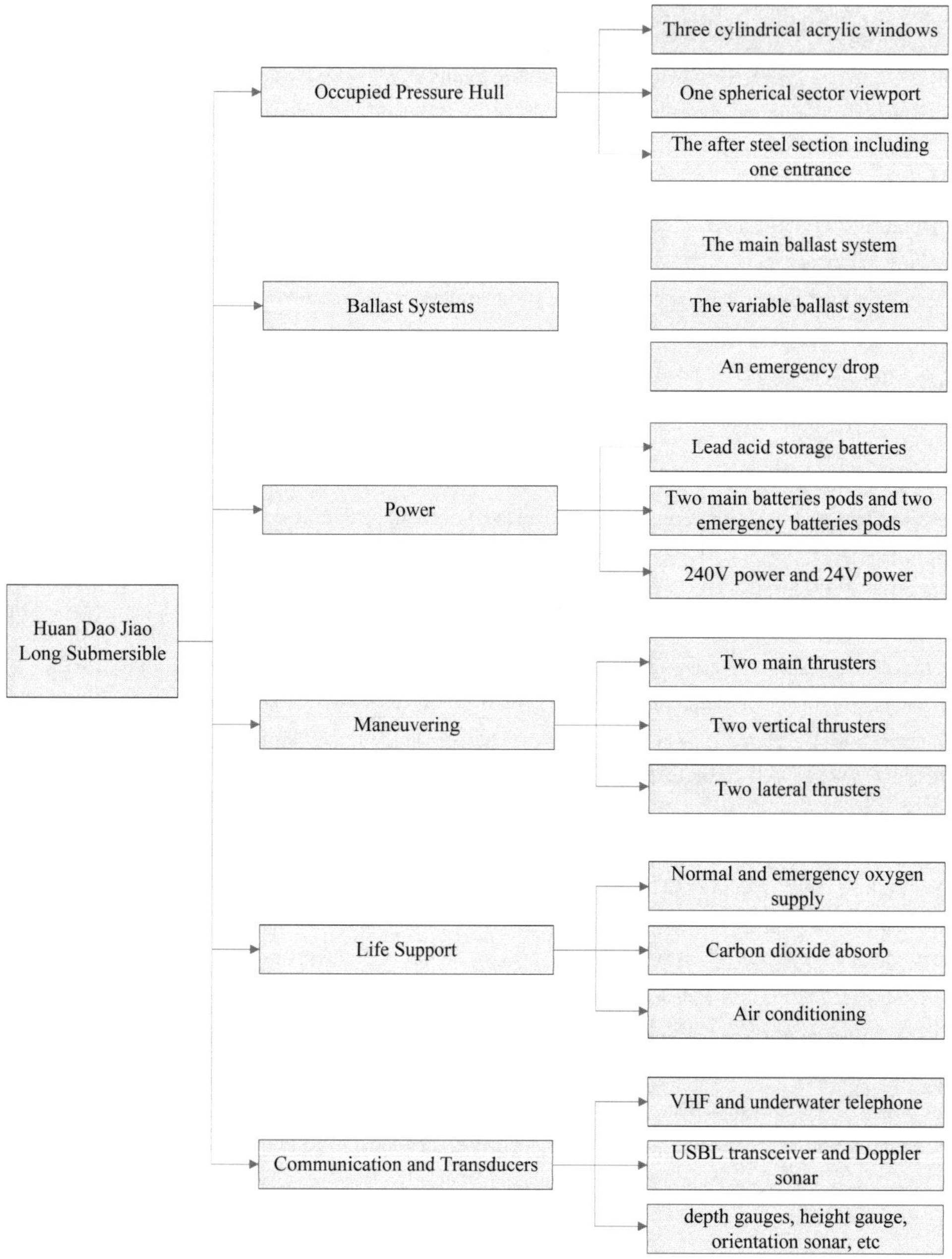

Figure 1 The main design considerations

The pressure hull is comprised of one spherical sector viewport and three cylindrical acrylic sections joined to an after steel section. Ring joining frames are placed between the cylindrical sections, while six tie rods pre-load the structure in the axial direction. The forward 150° hemispherical sector acrylic viewport for the passengers. The after steel section is composed of AH36 steel. One entrance and hatch are 650 mm in diameter.

The submersible has a main ballast tank system composed of a forward and an after tank. Decompression air was used to blow down the main ballast tanks when less than 10 meters depth.

The purpose of the main ballast tanks is to provide the submersible with the necessary freeboard, stability and buoyancy while in the surfaced condition. Main ballast tanks can also be blown at 40 meters with high pressure air in emergency condition, resulting in a rapid ascent.

The variable ballast tank system composed of a forward hard tank and an after hard tank, is designed to allow the submersible to be neutrally buoyant. The two variable ballast tanks are one-atmosphere, pressure resistant tank, forward and after tank longitudinal symmetrical located in the center of gravity. Also using decompression air to blow down the variable ballast tanks. Variable ballast tanks can also be blown at 40 meters with high pressure air in emergency condition.

An emergency steel drop weight is located under the center of gravity. The weight is released by the actuation of two hydraulic cylinders powered by two manual pump mounted in the pilot's compartment. Releasing the drop weight will allow the submersible to surface in emergency condition.

The submersible derives power from lead acid storage batteries. The batteries are contained in four external one-atmosphere battery pods. Two main batteries pods contain 240 V groups of cells and 24 V groups. Two emergency batteries pods contain 24 V emergency group cells. Power is sufficient for 8 hours of submerged travel at the average rate of consumption. The submersible has well designed electrical systems with extensive circuit protection and ground detection capability. All electrical equipments in human occupancy cabin use 24 V power.

The maximum speeds of the submersible is 3knots, cruise speed is about 0.5 knot. There are two main thrusters powered by 240 V groups of cells. The vertical thrusters are contained a forward vertical thruster and an after vertical thruster. The submersible dive with litter buoyancy, the vertical thruster are always work when the submersible submerge travel and dive. Two lateral thrusters is located forward and after respectively. Maneuvering is intuitively accomplished through a single proportional joystick control.

Life support system composed of normal oxygen supply, emergency oxygen supply and carbon dioxide absorb. High pressure oxygen is stored in individual bottles located outside the pressure hull. An independent emergency oxygen system consists of five bottles. The carbon dioxide component of the air in the passenger cabin is removed through soda-lime. Soda-lime was put in absorbent hull with fine screen. Fin under absorbent hull was used for increasing absorption effect. In order to save space, lithium hydroxide was used for emergency carbon dioxide absorbent. The normal and oxygen supply and normal carbon dioxide absorbent can meet requirements 12 people over 8-hour period and 72-hour period respectively. The oral-nasal masks located under the seat, one per person, it is capable of using over 30 minutes.
The submersible also have an air conditioning system to cool and dehumidify the cabin air. Cool air flow to the passenger through pipes. Fire protection includes fixed fire detection and fire alarm system, one fire blanket and three portable water base type fire extinguishers. Emergency food and fresh water, inflatable life jackets and first aid outfit are included in the human occupancy cabin.

VHF used for communication on the surface, and underwater telephone used for communication in the water. A forward and an after video camera provide external viewing to the pilot through a monitor. A forward and an after anti-collision sonar provide distance from obstacle. An Ultra-Short Baseline (USBL) transceiver was mounted on the submersible, provide depth and underwater moving track of submersible to the support ship. Other instrumentation includes two depth gauges, one height gauge, one orientation sonar, Doppler sonar, O_2, CO_2, H_2 concentration transducers, etc.

3 CONSTRUCTION

The purpose of the general assembly of the submersible is to assemble and integrate the pressure hull, equipment, pipes and cables into a complete submersible body system.

As is well know, the main difficulties in the general assembly are assembly of three acrylic cylinders. Many efforts have been made in order to assemble acrylic cylinder. We decided to use horizontal assembly plan and manufactured special auxiliary fixture. In order to overcome interference fit difficulty between gasket, cylinder window and O ring, we make the most of thermal expansion and contraction characteristics of acrylic window, we raise the temperature when assemble the hard bearing gasket, and reduce the temperature when assemble the O ring. In order to reduce the assembly difficulty, the hard bearing gasket and O ring were assembled separately.

After the assembly of spherical window and cylindrical windows, Vacuum test was conducted for verifying sealing property of human occupancy pressure vessel. The pressure in the vessel was below 10% atmospheric pressure, maintained more than 24 hours.

After the general assembly, we have conducted 49 items commissioning on land.

4 OPEN WATER TANK TEST

The tank tests of the submersible have been done at the open water tank of CSSRC. During open water tank, we measured the weight and displacement of submersible, we have successfully completed all 17 functioning tests, all the test results have met the designed requirements.

5 SEA TRIAL

The submersible was transported from Wuxi city to Sanya city in May 26th, 2016. In May 27th , 2016, the submersible was launched to the support ship, then finished battery charging, inflation high pressure air and inflation oxygen. We conducted the first diving in May 28th, 2016. The general operation procedure of the sea trial is composed of test preparation, land inspection, launching, surface inspection, project test, recovery and maintenance. All the test results have met the designed requirements. The list of sea trial and measurement items as well as the requirements to be reached are shown in Table 1:

6SUMMARY

Huan Dao Jiao Long , China's first acrylic-hulled submersible, is just our first try in tourist submersible. We also hop design and construct an bigger and more comfortable acrylic-hulled submersible.

Table 1 List of sea trial test and measurement items

No.	Item Name	Requirements	Sea Trial Contents
1	Balance Test	Diving with litter buoyancy	Change the water quantity of variable ballast to achieve balance between weight and buoyancy
2	Pressure Test	There are no leakage and destroy in pressure hull	16 , 24, 32, and 40 meters depth, separately maintained more than 1 hours
3	Power distribution test	Normal functional of all electric equipments	1. Supply power for all electric equipments 2. Measure hydrogen concentration of battery pods when battery work 3. Switch battery power
4	Ground detection test	1. Ground current lower 0.1 mA in land condition 2. Ground current lower 0.3 mA in surface and underwater condition	Detection ground current of 240V and 24V power supply circuit
5	Main ballast supply and draining test	Intake time < 300 seconds Drainage time < 100 seconds	1. Water intake and drainage tests of the main ballast tank in normal condition 2. Water intake and drainage tests in emergency condition
6	Variable ballast supply and draining test	Intake time < 300 seconds Drainage time < 60 seconds	1. Water intake and drainage tests of the variable ballast tank in normal condition 2. Water intake and drainage tests in emergency condition
7	Functional test of releasing emergency drop test	Normal functioning of releasing emergency drop	Release 1670 kilograms emergency drop
8	Functional test of the observation equipment	Normal functioning of all equipment	Functional tests of video cameras and underwater lights
9	Functional test of Acoustic equipment	Normal realization of functions	Functional tests of two anti-collision sonar, one height gauge and one Doppler sonar
10	Functional test of orientation sonar	Date is correct	Functional test of orientation sonar
11	Functional test of communications equipment	Clear talk	Inspecting VHF communication functions on the surface Inspecting underwater telephone

			communication functions in the water
12	Functional test of depth gauge	Dates are correct	Functional tests of two depth gauge
13	Date detection of transducers	Dates are correct	Communication functions between all transducers and computer
14	Performance test of the life support system	Oxygen concentration: 17% ~ 23% CO2 concentration < 0.5%	Normal oxygen supply, emergency oxygen supply, CO2 absorbing
15	Manual control test in normal mode	Functions can be realized Maximum speed: 3 knots	Forward and backward moving Descent and ascent moving Turning left and turning right Moving left and moving right Forward and backward leaning
16	Manual control test in emergency mode	Functions can be realized	Forward and backward moving Descent and ascent moving Moving left and moving right
17	Functional test of imaging sonar	Functions can be realized	Functional test of imaging sonar
18	Functional test of USBL	Display depth and underwater moving track of submersible	Functional test of USBL
19	Functional test of radar transponder	Display surface moving track of submersible	Functional test of radar transponder
20	Functional test of strobe lamp	Functions can be realized	Functional test of strobe lamp
21	Functional test of whole procedures	Normal functional of all equipments	No stop test for 8 hours

References

1. LIU Shuai, YE Cong, HU Zhen, ZHANG Xiaoxian; Status and Prospects of Tourist Submersibles [J]. Shipbuilding of China, 2013, 54 (03): 200-208. (in Chinese)
2. Jerry D. Stachiw. Handbook of acrylic for submersibles hyperbaric chambers and aquaria [M]. Best Publishing Company, 2003.
3. Bruce L, J. The DeepView 66: A brief technical description [DB/OL].2012-10-30]. http://www.ussubmarines.com/resources/Deep.66.pdf.
4. China Classification Society. Rules for Building and Classing of Diving system and submersible[S]. Beijing: China Communications Press; 1996. (in Chinese)
5. The American Society of Mechanical Engineers. Safety Standard for Pressure Vessels for Human Occupancy [S]. 2012.

A Vision of the Future of Manned Submersibles

By Graham Hawkes **HAWX Open Ocean Inc.** January 12, 2017

INTRODUCTION

The submersible business has changed dramatically in recent years. Shaped in the last century by undersea work needs of the offshore oil and gas industry and today by softer emerging recreational markets. The future of personal undersea recreational craft with safe and zero environmental impact is guiding the research and development of the author's latest projects. Based on available data and the author's fifty years experience, this paper reviews the basics for a new submersible business. The author is looking for potential revenue sufficient to attract venture capital. The foundational argument is that this is an ocean planet, where the oceans remain underutilized with unlimited room for growth.

A BRIEF HISTORY OF UNDERSEA MILESTONES

In the 1970's, enthusiasm for developing deep diving lead to the speculation that the future would see lungs flooded with incompressible fluid enabling humans to reach great depths. However, the discovery of successive physiological problems culminated with emergence of High-Pressure Nervous Syndrome (HPNS) effectively ended that research and hope. Divers at Comex laboratory in France set the ultimate pressure depth record at 2,290 feet of sea water (fsw) over thirty years ago, breathing mixed gases while encountering HPNS. While todays SCUBA standards typically sets the maximum depth at only 30 fsw without need for decompression. It is reasonable to conclude that these depth numbers represent the limits of human entry into the oceans, unless protected from depth-pressure. Therefore only manned submersible vehicles accommodating occupants at one atmosphere can meet the need for direct human access into the oceans at depths below practical diving.

In general, the major barrier to traditional manned submersibles is the high overhead cost of their mothership. Motherships were required to transport and launch traditional submersibles over their target dive site because the subs were slow with limited range. Historically, such ships need to be very large and specialized to safely handle the weight of a traditional submersible, today the most capable submersible support ships can cost up to $100,000 per day to use. This model continued unchallenged for decades.

In the 1980's the introduction of the unmanned Remotely Operated Vehicles (ROV) proved disruptive to the traditional mothership-submersible business model. The cost advantage was primarily that the ROV did not need such a large or specialized support vessel. Rather, ROV operations could be flexible and rent a suitable workboat on an as needed basis, known as a "ship of opportunity." And furthermore, for simple subsea observation, a low cost "eye-ball" ROV could be used without heavy tether management systems, deployed from smaller low cost support vessels. The inevitable outcome being that the traditional business of manned submersibles seemingly dried up overnight.

At the beginning of this new century, an unexpected new market opened, a bright spot in a depressed market. The emerging market was for privately owned and operated submersibles carried aboard very large private yachts. Such yachts are known today as Mega Yachts. Mega Yachts are sometimes defined as those costing over $70 million dollars, for example the SV Maltese Falcon (built for the late Tom Perkins). The 293-foot, groundbreaking Maltese Falcon carried the author's DeepFlight Super Falcon on its teak foredeck, which was launched by an electric hoist under the forward lower yardarm. [Note: the name Falcon was a coincidence. The author honored a commitment to name a flying submersible after Sir, Peter Falcon Scott.]

PART 1.
Current State of the Submersible Industry

Summary

Based on anecdotal submersible activity data provided by Will Kohnen, Underwater Intervention Conference Chair, it is clear that new builds of submersibles for the (private) experiential market already significantly out number new builds for traditional markets. Showing that the manned submersible business is transforming from its traditional role as subsea work system to recreational markets. It is worth noting that, Atlantis Submersibles by offering low price point ticket rides has exceptional high revenues for an existing submersible business.

Traditional Submersible

Traditional submersible for the purposes of this paper refers to the submersible type that is relatively unchanged since the 1960's. Their purpose was subsea work for government, science and industry. The traditional submersible was typically equipped for work with manipulators and was usually supported aboard its own dedicated mothership. An example of this model being the ALVIN operated from R/V Atlantis by Woods Hole Oceanographic Institution.

The surviving application for traditional submersibles are national undersea science programs of a few wealthy countries, China was the last to join this elite group. These typically larger and deeper HOVs operate with a pilot and two scientists. The surviving few are typically looking for new sources for operational income. While this research class is showing resilience the number of submersibles retiring outnumber the new builds shrinking an already small market.

Diesel Electric Submersibles

This newer larger type is differentiated from traditional submersibles by their surface running capability, usually powered by a snorkeling diesel engine. This type is designed to be launch from shore and operate with a limited range, requiring it to return to base. This method eliminates the overhead cost of a mothership, at least for near shore daily operations. Although there are a few different examples of this type of submersible typically equipped with manipulators for science applications, there are no indications that their numbers will expand substantially, with the possible exception of limited military or tourism applications.

Atmospheric Diving System/Suit (ADS)

This radically different type of working submersible may be thought of as a hard diving suit that maintains a person at one atmosphere thus allowing human access to greater depth. Note the author set a depth record in testing his ADS Wasp at 2000 fsw in Royal Navy test facility in the south of England 1975. Examples of early ADS are the Jim suit, named after its first test diver, Jim Jarrett and the authors Wasp and Mantis. The Wasp and Jim suit were commercially owned and operated for many years by Oceaneering International. All ADS of that era were tethered to provide power, communications and the means for launch and recovery. But ultimately competition from the ROV won and today ADS are almost extinct. An exception are the products from Phil Nytten who has continued innovating the type.

DIY (Personal) Submarines

Perhaps the best examples of personal submersibles were based or inspired by the Kittredge K-250 submarine. The DIY class is supported by enthusiasts who may build and operate submersibles outside industry standards. Their numbers fall below significant business levels. This class appears to be a perennial exception to any rule.

Acrylic Pressure Hulled HOV for the Mega Yacht Market

Notably the media serving Mega Yachts from about 2005 began reporting that the new "toy" of choice was a submersible, although in fact very few yachts own them. Some companies saw the opportunity and designed lighter and more attractive modern acrylic submersibles. Generally, they were built with standard PVHO acrylic hulls to afford attractive visibility. Most commonly the pressure hulls are supplied by Stanley Plastics in the UK with certified depth ratings of 3,000 fsw or less. Examples of this type of construction are those from SEAmagine Hydrospace Corporation, with innovative surface buoyancy systems improving safety and ease of use. A new company in this field is U-Boat Worx who have introduced several smaller modern submersible designs that have won the biggest market share of this emerging market. Note, this type of submersible usually does not have manipulators although as with most rules there are exceptions.

There are reportedly over 7,000 Mega Yachts with about 80 new constructions per year. However only about the largest 10% are big enough to accommodate the modern 13,000-pound submersible, reducing the actual market to a few. The actual cost of the submersible does not seem to be an issue for this market. Rather it seems that the limit to sales is the small size of the actual market and high barriers to use without additional justifying utility. For example helicopters are also a "toy" but can be further justified as transport utility.

New Submersible Types for the Mega Yacht Market

The author designed the DeepFlight series starting with DF1 in 1996 as a wholly new type of submersible firstly to explore underwater flight. That was then further developed

through the Aviator built at Sub Sea Systems to the Super Falcon for the mega yacht market. The Deepflight series used ADS type pressure hulls to reduce the weight by half. The next new type designed by the author was the DeepFlight Dragon again built by Hawkes Ocean Technologies, created for a smaller footprint on deck.

But this was not enough to keep up with changing yacht designs. To create a sleeker yacht, yachts are now designed with internal garages for "toy" storage. Therefore, submersibles could no longer be housed on an open deck, but need to compete for storage space in an even more restricted below decks garage. The "toy garage" typically opened to the stern or side. The "toys" then launched by overhead horizontal rail crane extending out of the garage bay. Generally the garage is designed to accommodate the owner's launch, jet skis and other small craft. Hence the author's designs were as compact as possible with removable sections to fit within the garage space. The most limiting dimension was usually internal height of the garage under the overhead crane. Another response to too many toys has been to have a second utility vessel accompany the main yacht, these were called "shadow boats." The shadow boat then became the obvious home base for a submersible. But again the market size is very small.

PART 2.
Emerging Markets With Future Potential Growth

Summary

Indicators suggest that manned undersea craft will move almost entirely from undersea work markets to personal recreation with ticket and adventure markets leading revenue projections. Several builders such as SEAmagine Hydrospace Corporation are already addressing commercial recreation ventures with larger submersibles. In addition a partnership between Rainbowfish and Hawkes Ocean Technologies has announced a new experiential venture due to start directly serving guests at ocean resorts in 2017.

Submarine Rides and Ticket Sales (low price point undersea experience)

Here is a general assumption that the lower the price point the greater the revenue. This would indicate that the biggest business is in the lowest cost under sea ride/ticket. This seems to be demonstrated by Atlantis Submarines, who have taken many millions of people on undersea rides for approximately $100 per person.

Others have offered ticketed ride experiences, but at higher price point such as Stanley Submersibles of Roatan, Honduras. The higher cost business model is usually an experience to greater depths with fewer people. This price range is comparable to helicopter adventure rides that range from $300-$500 dollars. However, todays total numbers for submersibles operating and selling more expensive rides falls below investment business significance. On the other hand, the helicopter comparable suggest this could be a strong future business.

Adventure Experience

Adventure involves a longer and more immersive experience than the lower cost ticketed ride. A notable comparable for the adventure experience is Virgin Galactic, claiming several hundred thousand customers each paying $100,000 dollars. Virgin Galactic is an indicator that the adventure business can be significant. While the $20-$40 million charged the seven space tourist by the Russian Space Agency represents the extreme high end.

It is worth noting that a few enterprises have run one off adventures in the deep ocean. Deep Ocean Expeditions charged $60,000 dollars to visit the Titanic. The author has run occasional three day "flight schools" charging $15,000 per student. Again suggest this could be a strong future business.

PART 3.
Major Factors Affecting the Future Submersible Industry

Future Challenge to Low Cost Ticket Sales

Virtual Reality (VR) is going to be a major competitor to the low cost subsea ticket market. It is conceivable that ROV's with VR camera systems will dive into the deep ocean with customers enjoying the experience via VR at home. That said, the authentic first person experience will remain the most valued experience likely unchallenged by VR.

Breath-Hold Diving
Given we all are naturally able to breath-hold beneath the sea as mammals, such simple and safe but brief diving may be an underutilized method with potential for future cost effective undersea experience. One example of innovative use of human ability to breath-hold is Sea Bob, an unusually powerful swimmer propulsion devise that is already a standard, $12,000 toy on many large yachts.

Regulations of Manned Submersibles
Safety is the major concern for regulation of submersibles. The industry can be expected to move beyond simple reliance on "certification" standards to "outcome based" rules. Modern rules are more flexible and less likely to lead to unintended consequences as future designs change. As designs change and move outside the protection of existing design based regulation, companies will need to assume a more proactive role in safety.

Mitigating Human Error
Atlantis Submersibles and Sub Sea Systems both have exemplar safety records, each having taken millions of guests safely beneath the sea. They both seem to have created an operational and training culture that focuses on eliminating human error as well as mitigating hardware malfunctions. The area of addressing human operational error is familiar to aviation and will be a new focus for the submersible industry as it seeks to serve increasing numbers of the public.

Regulation for the Environment
Regulations already outlaw small boat anchoring in some protected areas to prevent environmental damage caused by anchors. It is likely that future regulations will protect the environmental from submersibles contacting the ocean floor. Other regulatory restrictions anticipated include prohibiting submersibles with manipulators from diving protected shipwreck sites to prevent removal of artifacts.

Regulation for Pilots (Training)
The author is arguing for pilots of private subs not to follow the traditional submersible model, where the builder issues pilots licenses because this license is inadequate for the inexperienced. The first Wasp pilot licenses issued personally by the author are an example that this traditional model no longer applies. Those Wasp students were employees of Oceaneering and were already qualified diving professionals. The author was simply training to them operate a new type of equipment. When comparing this to aviation practice, the licenses issued would more accurately be "Type Training" certification. This would mean a qualified pilot was now training and qualifying on a different aircraft. As a solution, the author is proposing a multi-part licensing for future members of the public (non-ocean professionals) to become submersible pilots. The proposed licensing program is more comprehensive than the traditional licensing while at the same time having a relatively easy entry point.

Training Program
Part 1. Classroom and/or online instruction with some test establishing general theory and risk education for submersibles designed for an individual with no prior

experience. This is the equivalent to "ground school" in aviation as the first step to actual flight lessons. This qualifies another wise inexperienced individual to enter Part 2.

Part 2. Existing commercial submersible manufacturers training. This traditional manufacturers pilot training will reward the student with similar pilots license as previously but restricted to the submersible type. Part 2 combined with Part 1 is therefore a basic submersible pilots license.

Part 3. This advanced level would require a pilot to maintain a logbook to document "log time," time spent in control of a submersible type. This is equivalent to required minimum time spent flying for aviation as needed to upgrade from private to commercial pilots license for example. Log book minimum requirements could be set by regulators, operators or by insurance.

CONCLUSION

Because—unmanned vehicle types will likely be preferred whenever human presence is not essential— the future market for the submersible will be almost exclusively be for human experience or any other purpose where direct human presence is needed. Therefore the future business of manned submersibles will focus on the one attribute that unmanned systems cannot serve, first person experience. Only with one atmosphere accommodation in manned submersible vehicles can humans access the ocean at depth beyond practical physiological limits.

Until the volume of public participation justifies government investment in new safety and environmental regulation, the industry will need to be self-governing as it moves beyond traditional guidelines. Note minimum environmental impact will become more important as the public understand we are opening a new planetary frontier to human access.

The principle future submersible business will be for safer and environmentally sensitive operations deriving revenues from providing human access into the oceans. Subsea rides returning to point of origin will likely earn the greatest revenues. However ticket rides will face completion from VR products for mass subsea experience. While more immersive and costly (adventure) experiences will be the most valued extending the business away from competion by VR. It is clear to the author that ocean space is our last mysterious frontier with potential for unlimited growth. It will also fuel curiosity and desire for adventure. Rewarding the emerging business that are focused on public access using submersible.

Window Seat Weight Reduction Exploration with Non-Traditional Seat Geometry

Thomas Trudel[1] and Karl Stanley[2]

[1]Lead R&D Engineer at OutsideInnovation and Stanley Submarines, Roatán, Honduras
[2]Owner and Pilot at Stanley Submarines, Roatán, Honduras

***Abstract* —This paper provides an overview of the design and feasibility testing of a non-traditional submersible window seat design. Typically, window seat geometry is guided by the current revision of the ASME-PVHO-1 engineering standard as well as other references by the ASME, Stachiw, etc. As viewing area increases, window seat geometry is driven by the size of the acrylic window and not by the requirements for a hull penetration of equivalent size. The discrepancy in strength between the submersible hull materials and the weaker acrylic window can result in a window seat that is overbuilt for the required hull integrity and cause weight budget penalties as viewing area increases. The research presented in this paper focuses on non-traditional window seat geometries that decrease weight while performing comparably, in some respects, to designs that conform to the current ASME-PVHO-1 design standards. A new type of window seat is proposed with reductions in window seat weight between 22% and 33% at the potential expense of a shorter window service life. Design methodology, assumptions, FEA results, deviations from the standard, and empirical design studies are summarized in detail. Two model flanges and windows designed using this new methodology were built and tested. One of the designs was tested in the field near its design depth on 102 dives where it showed signs of wear but did not fail. FEA constraints were validated using strain gauge and displacement measurements on the conical and low pressure faces of the windows. STCP is scheduled to be evaluated in a hydrostatic pressure chamber prior to the UI 2017 conference presentation.**

I. Introduction

During the design of viewports for deep sea applications it quickly becomes apparent that large viewports can cause significant weight penalties due to the size of the acrylic window and required seat to support it. A typical hull penetration requires that 100% of the weight of hull removed be added to the geometry of the penetration, or in the case of acrylic viewport penetrations slightly less than 100%[1]. However, because of the strength discrepancy between window materials (likely acrylic) and hull materials (likely steel or titanium) only a narrow selection of window sizes will allow the window flange to equal the weight of the removed hull segment. As the internal diameter of the window increases for a fixed hull diameter, the amount of additional weight added to the submersible beyond the minimum for an equivalent hull penetration also grows.

Most deep diving submersible designs mitigate this problem by choosing windows that have a sufficiently small internal diameter such that their thickness is marginally larger than the hull thickness of the submersible. This decreases the excess window seat weight that must be added to support the window. Still hull thickness can become large near viewports growing to 6 inches on the Bathyscape Trieste[2] and greater than 4 inches on the original Alvin sub[3].

Window design geometry is primarily driven by design depth, internal diameter of the window, and t/Di ratios derived from the ASME-PVHO-1 standard. This makes the size and weight of the window seat a dependent variable in the system as well as a promising starting point when considering optimizing the design.

The research presented in this paper was initiated during a separate design study that was focused on increasing viewing area. Numerous methods were investigated with FEA and other analytical tools. One of those methods was to decrease the system weight required for a viewport thereby allowing a larger window to be included for the same weight penalty. This paper focuses on one of those potential methods to decrease the weight of a submersible window seat.

II. IDENTIFYING SYSTEM OPPORTUNITY

Early in the project, an attempt was made to investigate exactly how much additional weight was required for window seats as their thickness grows beyond the hull thickness. An initial study was conducted to characterize the excess window seat weight that would be required for a fixed window ID and fixed hull ID at increasing design depth. The intent of this study was to determine if there was enough potential weight savings available to justify investing in the project moving forward.

The study was structured in a simplified manner. The authors sought to design a set of minimalized window seat models for a 5.75" ID window at design depths ranging from 1000 ft – 36,200 ft. This set of window seats was then used to calculate the 'minimum excess window seat weight', which is herein defined as the amount of window seat weight in excess of the hull weight removed to accommodate the window seat. This would allow for the excess window seat weight to be defined as a function of design depth and help to identify the potential opportunity associated with an optimized window seat design.

The study boundary conditions were as follows:

1. Window ID fixed at 5.75" (window seat Df derived from ASME-PVHO-1)
2. Hull ID fixed at 48" (hull thickness determined using UnderPressure and FEA)
3. Window design conforms to current revision of ASME-PVHO-1
4. Window Seat shows initial signs of yield at design depth
   * This condition was used to provide a clear indicator that the designs were near their minimum potential size
5. Window seats show similar stress and displacement distributions at varying design depth
6. Displacement no greater than 0.1 inches in radius (relative to centroid of window seat)
7. Displacement gradient in conical bearing surface no greater than 25%
8. Displays similar displacement distribution at varying design depth

Using these criteria, window seat geometries and associated FEA results were generated to determine the minimum additional window seat weight for each seat. The results from the FEA study can be seen in Figure 1.

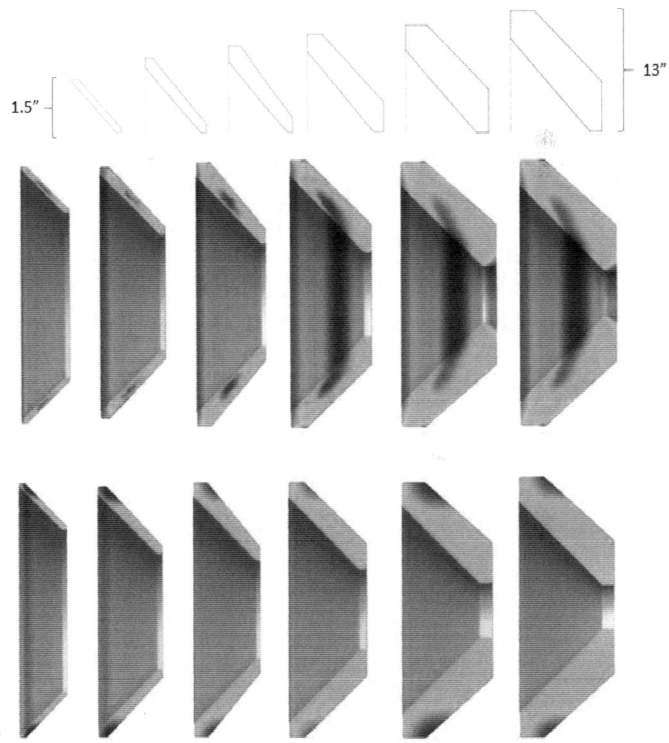

Figure 1. FEA stress (top) and displacement (bottom) comparison of the minimalized window seats used for the design study. Design depth increases from left (1000 ft) to right (36200 ft). It is clear that the window seats perform in a similar manner at each design depth. Maximum (red) on each FEA scale corresponds to 80,000 psi and 0.01 in, respectively.

Next, hull thicknesses were generated using the Deep Sea Power and Light, UnderPressure program and verified using FEA. Excess window seat weight was then calculated at each design depth using the hull models and the window seats from Figure 1. The results are summarized in Figure 2.

numerous precedents for using honeycomb structures in this way ranging from aerospace composites[6] to layers in an alpine ski or similar laminate structure.

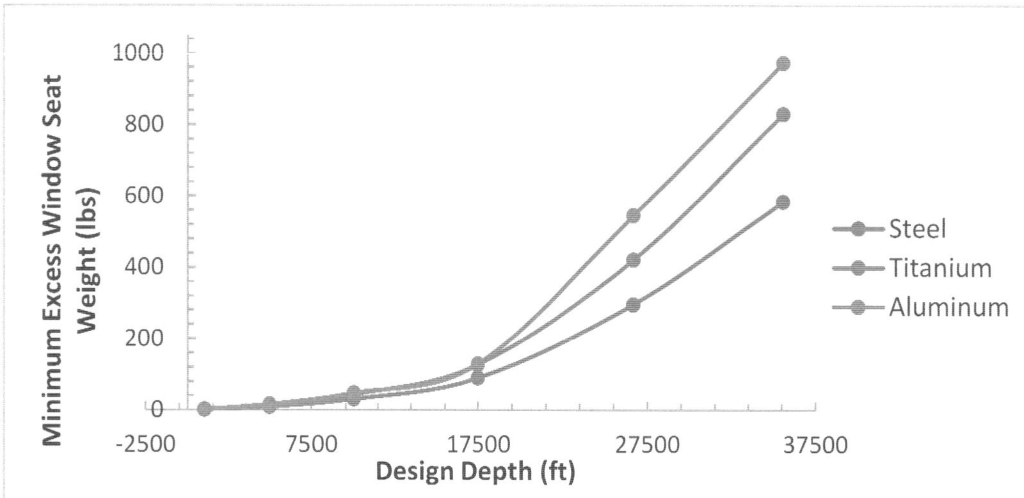

Figure 2. Design depth versus minimum excess window seat weight (window ID fixed at 5.75 inches, hull ID fixed at 48 inches).

The results in Figure 2 clearly show that the amount of excess window seat weight increases with design depth and the greatest opportunity is in vehicles that operate below 10000 ft of depth. The study also showed steel to be the most efficient window seat material due to its higher elastic and flexural moduli compared to titanium or aluminum.

III. WINDOW SEAT WEIGHT REDUCTION EXPLORATION - DESIGNS AND FEA

In this section three new window seat designs are presented that significantly decrease the weight of the seat while still satisfying the performance conditions from the excess window seat weight study.

When designing a window seat for submersible applications stiffness and yield strength are two of the primary factors to consider.

As an initial experiment in removing window seat material and preserving stiffness using a honeycomb structure, seemed like a logical starting point. Holes perpendicular to the conical bearing seat were used to create the honeycomb due to their simplicity and ideal stress distribution[4,2]. Additionally, a cylinder (hole) is stiffer than a solid bar of the same weight due to its area moment of inertia[5]. There are

When placing holes in the window seat determining the ideal packing density quickly becomes an area of interest. Without knowing the maximum allowable hole size at each location on the conical bearing surface, determining the ideal distribution is not possible. Also, the additional stress caused by the hole is partly a function of the hole diameter[7] and one would expect there to be a theoretical limit below which the holes no longer affect window performance.

Each of the three window seats followed the same design methodology but varied the size and location of the holes to start trying to define the boundary conditions for the potential acceptable hole sizes. Each design was characterized with FEA and was required to satisfy the same standards for stiffness and stress distribution as those window seat designs used in the excess window seat weight design study. The following section details each of the 3 designs and shows the FEA results at their design depth. The potential weight savings for each window are also listed.

As a baseline, a traditional window and window seat for the same design depth was created.

Figure 3 identifies the dimensions of the baseline window used in the study. Windows 1, 2, and 3 are all similar in size, but they did increase in diameter compared to the baseline in order to maintain performance when including the holes. The size increase was negligible as is evidenced by the weight savings achieved in the 3 window concepts.

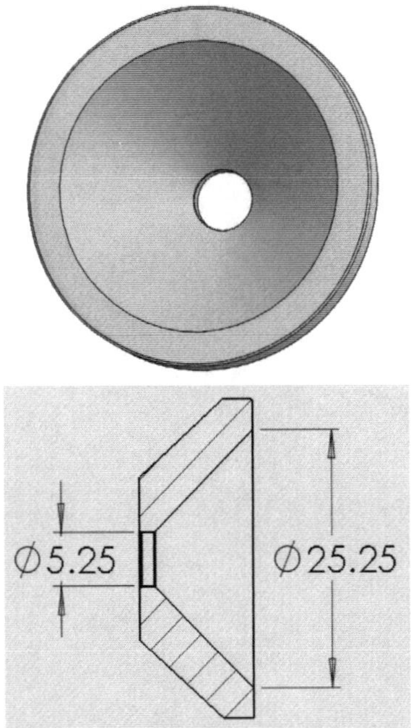

Figure 3. Baseline window model and dimensions. Design depth is 36,200 ft.

Table 1 identifies design parameters and shows a comparison of the window seat volume and excess window seat weight for the concepts.

Figure 4 shows a side by side comparison of FEA for all 4 windows (baseline and 3 experimental concepts) at their design depth.

Window 1 attempts to minimize the size of the holes to a limit where the simulation no longer shows any signs of yield in the window.

Window 2 attempts to use the largest holes possible in the window seat. Design criteria included using the largest holes possible and to show stresses in the window consistent with those seen at the conical edge of the low pressure face.

Window 3 attempts to place holes in the seat that were in between the hole sizes in Windows 1 and 2. The condition that the holes must maintain a window t/D ratio greater than or equal to 1 (normal to the hole) was enforced with this design. The holes nearest the low pressure face of the window are reduced in diameter to increase the stiffness and decrease stress in this region of the window flange.

Table 1. Window design parameters, window volume and excess window weight comparison. The first 4 parameters are the same for all windows.

Design Parameter	Baseline	Window 1	Window 2	Window 3
Design Depth (ft)	36,000	-	-	-
t/Di	1.7	-	-	-
Di (in)	5.78	-	-	-
Df (in)	5.25	-	-	-
Seat Volume (in^3)	4036	3520	3346	3410
Excess Seat Weight (lbm)	584	448	390	416
Decrease from Baseline (%)	0	23.2	33.2	28.8

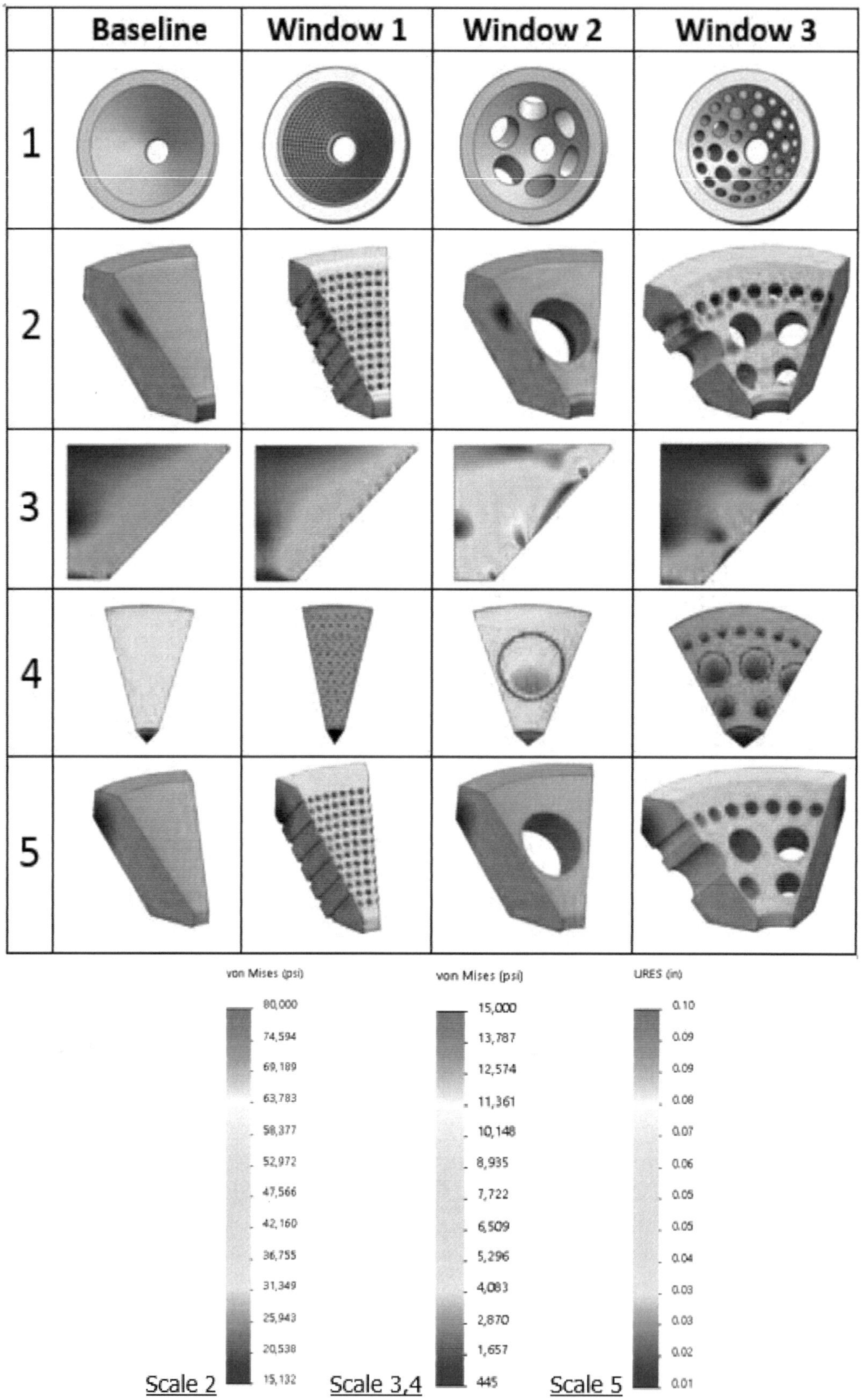

Figure 4. FEA result comparison for experimental window seats.

It may occur to readers that the designs presented above could be improved simply by placing a layer of material of minimal thickness at the interface between the seat and window (see Figure 5). Additionally, many hollow shapes could be used for the hull material removal (i.e. cylinder, box, truss structure, etc.).

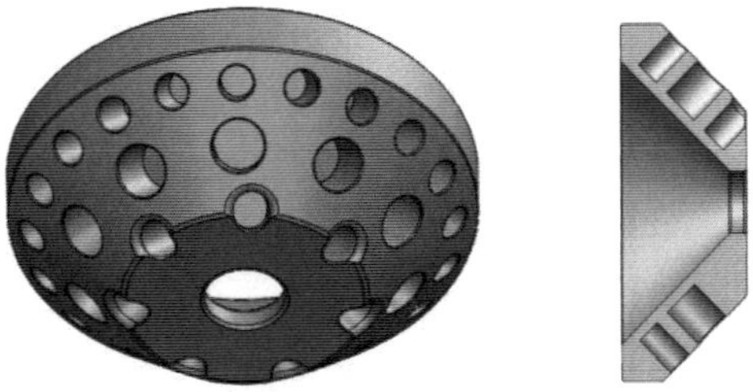

Figure 5. Potential window with removed weight, but continuous window bearing surface.

This fact was not lost on the authors. The window seats chosen for the initial study were motivated by several factors:

1. If a window can reliably be paired with a window seat of designs 1, 2, or 3, then it would be possible for existing window seats to be modified this way.
2. The authors were attempting to use a fast fail methodology and conserve funding so they decided to test a more extreme concept.
3. Window 1 showed favorable performance in the FEA for both the window and window seat. The factor of safety went down by 15-18% compared to a traditional window seat, but the window seat exhibits minimal performance penalties.
4. The authors were most curious if the acrylic window could survive using this type of window seat

IV. Experimental Campaign

Windows 2 and 3 from the initial design study were built as an empirical test to compare to the FEA results and current engineering standard. Window 1 was excluded from empirical testing at this time due to its performance in the FEA study and cost restrictions on the study. Windows 2 and 3 were chosen to explore the behavior of these types of windows and to attempt to characterize their limitations by exploring the more extreme geometries.

To better correlate the empirical test to a realistic window and window seat design the window baseline for this testing was copied from Idabel's pilot sphere windows. The design was scaled down by a factor of 2 for testing.

A drawing and pictures of the scaled down models of window seats 2 and 3 can be seen in Figure 6. Figure 7 shows stress and displacement FEA results for each window. The window seat FEA was excluded here for brevity, but conformed to the performance constraints of the other window seats in this study.

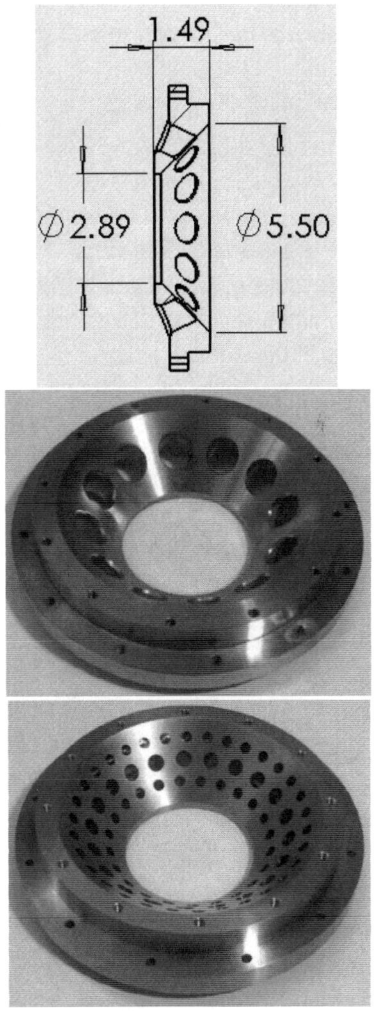

Figure 6. Scale model window dimensions and pictures.

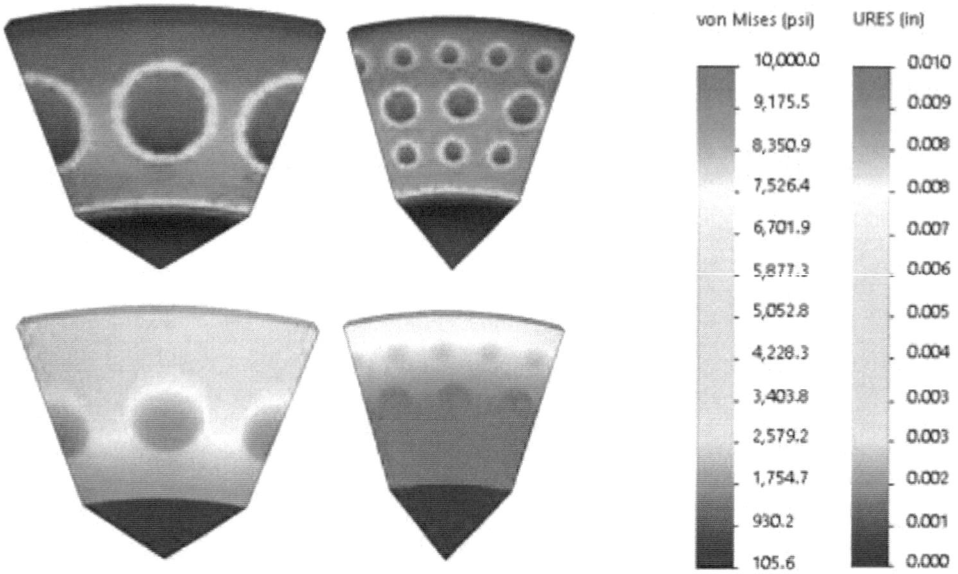

Figure 7. FEA of scale model Windows 1 and 2 at their design pressure (1800 psi). Top figures are stress, bottom figures are displacement. Note the change in scale for each plot compared to previous results.

The first test was conducted by cyclically loading the model Window 2 while mounted externally on the submersible Idabel, owned and operated by Stanley Submarines on the island of Roatán, Honduras. This approach allowed the authors to test the highest risk window design while conserving resources. Idabel's design depth is 2500 feet.

Cost constraints on the project prevented both window seats from being constructed simultaneously. Window 2 and its matching seat were built first along with a pressure vessel. Together they served as a camera housing while on Idabel for 102 dives during 2014.

After a year riding on Idabel to an average depth of 1500 feet the window showed little to no signs of crack initiation or adverse wear. There were two faint markings on the window that were clearly in the shape of the window seat holes, but they were sufficiently faint that they were difficult to see at the wrong angle and no change in the window seat shape could be measured, felt, or seen with a microscope (see Figure 8).

The design depth of the window (4000 ft) is more than double the average dive depth of Idabel, however, this result was still unexpected and

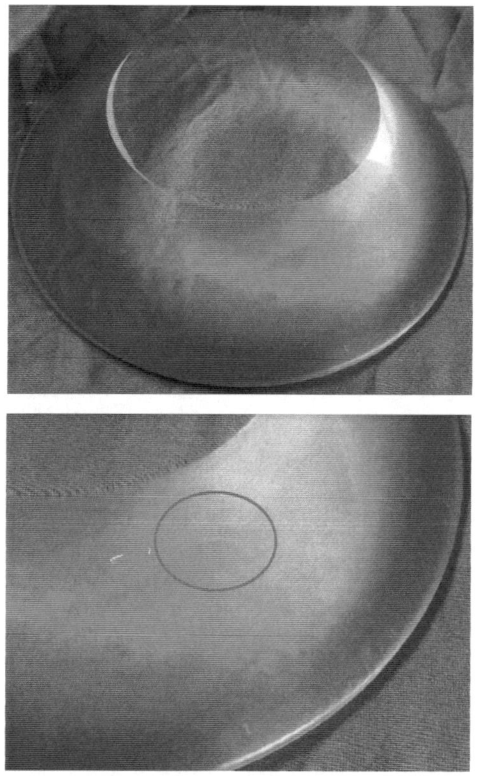

Figure 8. Window 2 after 102 dives to an average pressure of 690 psi (1500 ft). Notice the faint marks in the shape of the window holes.

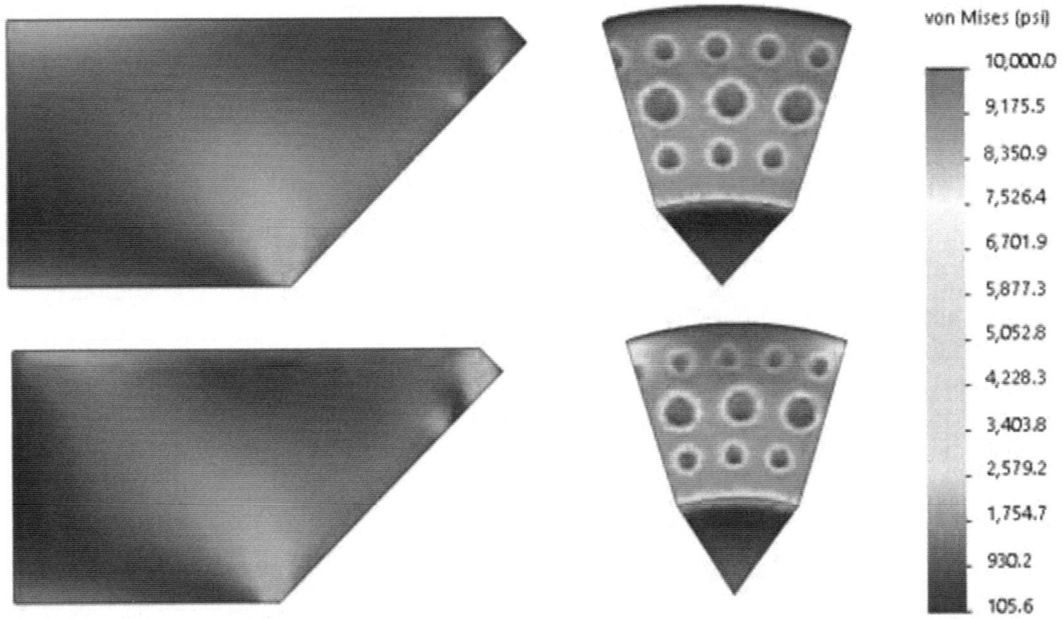

Figure 9. Hydraulic press FEA (top) versus hydrostatic loading FEA (bottom) at 1800 psi for Window 3. In the hydraulic press loading, note the decrease in stress at the edge of the low pressure face and increase in stress on the conical face. This is expected given the constraints of the the experimental setup. The same analysis was completed for window 2, but excluded here for brevity.

additional testing was performed to investigate the window design further.

It is clear from the FEA results that the acrylic window is the highest risk part in this assembly so most of the empirical testing focused on better understanding the performance of the window. Strain gauge and displacement testing were conducted using a 40 ton hydraulic press to simulate hydrostatic pressure at the intended design depth. Although this is not the ideal method to perform this type of testing, hydraulic presses have been used to successfully simulate hydrostatic loading in the hydroforming of metals.[8] Additionally, the hydraulic press testing did not rely on the experiment perfectly replicating hydrostatic pressure. The hydraulic testing helped verify FEA constraints prior to investing in hydrostatic pressure testing to determine the STCP of the test windows. Also, the windows were loaded to their design pressure and not to failure, so the non-uniform window displacement normal to the high pressure face was minimized. The primary goals of the cyclic testing on Idabel and with the hydraulic press were to put additional cycles on the windows near their full design pressure, look for signs of crack initiation or wear, and to compare strain and displacement measurements to the FEA.

FEA specific to the experimental setup was completed using the same constraints used for the other design studies in this paper. A comparison of the window stresses from a window under hydrostatic loading versus a window in the experimental setup can be seen in Figure 9. Herein, results comparisons between empirical and theoretical values use the hydraulic press FEA results.

The window seat was lubricated using a commercially available, silicone based lubricant. A 3 in thick cylinder of UHMW with the same diameter as the window was placed on the high pressure face of the window and the interface between the two was also lubricated. The UHMW cylinder was surface ground to a flatness better than 0.0005" to help ensure even loading. On top of the UHMW block were placed two, 0.25 in thick steel sheets that had been machined to a flatness better than 0.001". Last, a hydraulic pressure gauge was placed beneath the hydraulic ram to ensure repeatable force was delivered during each cycle.

A dial indicator with an accuracy of +/- 0.0005" was used to measure the window displacement into the window flange holes. The results of these measurements compared to the FEA predictions can be seen in Table 2 and 3.

A stacked strain gauge rosette was used to measure strain in the same location (see Figure 10). Strain readings were taken every 10 cycles from 0-100 total cycles. Predicted average strain values at the gauge location were determined by averaging the strain measurements from each node in the FEA within the gauge footprint. The analysis used for comparison was designed to reflect the setup in the hydraulic press as opposed to traditional hydrostatic loading. The results of these measurements compared to the FEA predictions can be seen in Table 2 and 3.

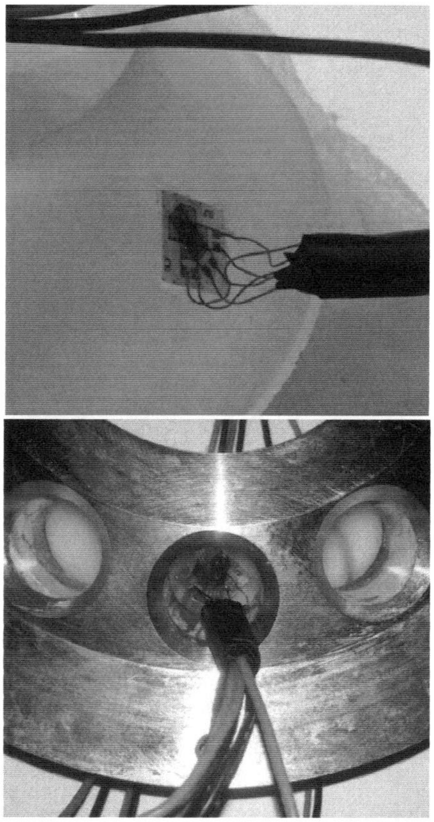

Figure 10. Strain gauge rosette mounted to window.

Table 2. Window 2 theoretical versus empirical strain and displacement results. Theoretical values were 0.0021 and 0.0021 in, respectively

Cycle Count	Strain Empirical	Disp Empirical (in)
10	0.0017	0.00125
20	0.0017	0.00125
30	0.0017	0.00125
40	0.0017	0.0015
50	0.0016	0.0015
60	0.0018	0.00125
70	0.0017	0.0015
80	0.0018	0.0015
90	0.0017	0.0015
100	0.0017	0.0015

Table 3. Window 3 theoretical versus empirical strain and displacement results. Theoretical values were 0.0019 and 0.0016 in, respectively.

Cycle Count	Strain Empirical	Disp Empirical (in)
10	0.0014	0.001
20	0.0014	0.001
30	0.0015	0.001
40	0.0015	0.001
50	0.0016	0.001
60	0.0016	0.001
70	0.0015	0.00125
80	0.0015	0.00125
90	0.0015	0.00125
100	0.0015	0.00125

All strain values recorded during loading in the hydraulic press were within 24% of the predicted value.

All displacement values were within 40% of the predicted value. It is also worth noting that the difference in individual displacement measurements was within the error of the measuring device. The resolution on this test was not high enough to heavily leverage the displacement results, but still serves as a relevant check of the FEA constraints.

The conical face of each window was examined after testing for surface defects or other signs of wear.

On Window 2 there were signs of wear in the shape of the window seat holes (see Figure 11). The marks could be felt when running a straight edge over them, but no change in height between the area inside and outside the marks could be reliably measured using a digital micrometer or caliper. Regardless, it is likely that these marks are signs of cyclic wear in the window. They might be localized yielding in the material near the conical surface or they could be the early stages of crack initiation. It was difficult to identify a cause given the size of the marks.

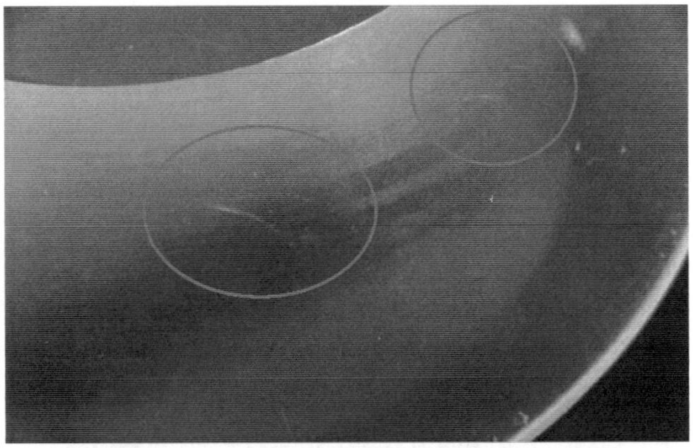

Figure 11. Marks on conical surface of Window 2 after 102 pressure cycles to average of 690 psi and 100 cycles to 1800 psi.

Window 3 showed no signs of adverse wear on its conical seat. There were very few marks on the window and none that matched the holes in the window seat (see Figure 12).

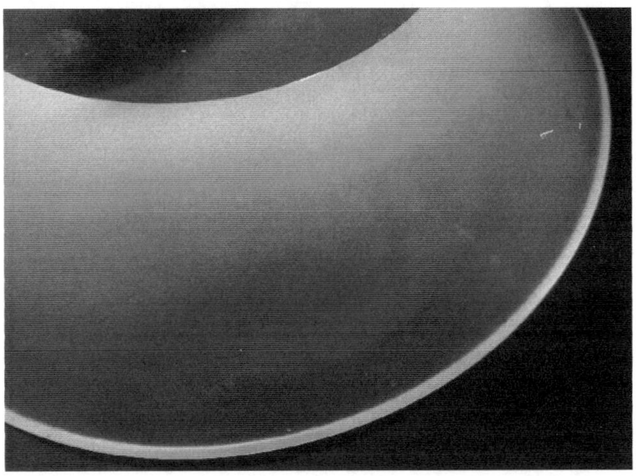

Figure 12. Window 3 conical surface devoid of wear marks after 102 pressure cycles to 690 psi and 100 cycles to 1800 psi.

V. Discussion

Based on these results, it is possible that Window 2 will incur penalties to STCP and window service life due to adverse wear in the conical window face leading to early failure. Window 3 has the potential to produce an acceptable service life and STCP value.

Based on the comparison between the theoretical and empirical values in Tables 2 and 3, it seems likely that the FEA constraints are realistic. Additionally, the performance of the window in the author's FEA showed comparable performance to published results of windows under similar loading.[9]

It seems as if there is potentially an acceptable hole size that could be repetitively removed from a conical window seat without degrading the short term performance of the window. The testing in this paper indicates that there may be an inverse relationship between hole size and performance. At first glance this may seem in contrast to tests that produced premature failure in window seats with annular o-ring grooves in the conical bearing surface or statements from trusted references that indicate small defects in the conical bearing surface may lead to failure.[10] However, on closer examination the same literature indicates that published data on the initiation of cracks on the bearing surface of the window are very limited and that the appearance of surface cracks precedes catastrophic failure by hundreds or even thousands of cycles.[11,12] Furthermore, the exact mechanism that causes crack initiation and propagation in PMMA under cyclic loading is not yet fully understood[13]. This seems to reinforce the potential for reducing weight by removing weight from the bearing surface of the window seat provided that a potentially lower service life (and higher operating cost) is acceptable.

By removing weight from the window seat and maintaining stiffness using a honeycomb structure, these results indicate it may be possible to decrease the weight of the window seat without incurring unacceptable performance penalties. Obviously, to produce an usable design with this methodology it would have to be fully qualified through extensive experimental testing to ensure safety and reliability metrics.

The results of this study could also be interpreted as additional evidence to support the conclusion that the ASME-PVHO-1 standard for submersible windows is conservative[14] and there may be room to improve the acrylic window standard and allow for expanded design criteria.

Regardless, the study produced an interesting result and hopefully serves as an additional data point in what acrylic submersible windows are capable of in extreme cases.

After this testing was completed, the same windows and flanges were scheduled to be tested to their STCP in a hydrostatic pressure chamber at Deep Sea Power and Light. That test is being prepared as of the writing of this paper.

VI. Sysytems Engineering Notes

When considering the practical application of a window of this type several advantages and risks come to light.

Removing window seat weight using this methodology could result in window seat weight reduction of 20% or more. In a window designed for full ocean depth that means a single window seat could stand to lose several hundred pounds of weight depending on the Di of the window.

When designing any weight sensitive system it is advantageous to identify areas where weight can be removed from the system and at what cost. For example, if it was shown that window seats of this type could save weight at the expense of window service life, it might be acceptable to many designers to replace the window more frequently and take advantage of the weight savings to include a larger window, additional batteries, etc.

Machining of a window seat of this type is cheaper and more reasonable than ever. The risk of defects from the machine process producing a failure has also dropped dramatically since it was identified as a risk by Stachiw during his early research[10]. Finally, the cost of complex CNC machining has plummeted in recent years so the additional production cost and risk of implementing a window seat of this type is minimal.

The idea of collecting real time telemetry from a submersible is also more realistic than ever. The cost of sensors and data acquisition equipment is decreasing exponentially and the quality of both systems is constantly improving. Strain, displacement, and other data could be collected real-time to monitor the performance of a window and warn the operator of potential signs of failure. Implementing an early warning system of this type could be critical to the operational success of systems that have a lower factor of safety.

Further advantage is gained when one considers that any conical bearing surface could benefit from this type of modification. Assuming that stiffness is a driving factor and that one could show that the honeycomb structure does not endanger the acrylic performance, it would seem that this methodology is applicable to a wide range of window designs (including semi-spherical window seats).

VII. Conclusions

As an exploratory study in its most early stages, there is only so much that can be determined from this initial study. The low sample size and lack of sophisticated test equipment limit the specific conclusions that can be drawn. Nevertheless, valuable information was gained.

As an exploratory exercise the study was quite successful.

The excess window seat weight design study characterized one design case and demonstrated the system level opportunity for weight optimization in this area.

It was demonstrated that it may be feasible to remove weight from the conical bearing surface of a submersible window seat using a honeycomb geometry without severely degrading the performance of the window it accommodates.

3 theoretical window seat designs were derived from those used in the excess window seat weight study. These designs successfully showed that the weight of the seat could be reduced by incorporating a honeycomb structure while maintaining comparable performance in window seat stress and displacement. The tradeoff was removing weight

from the conical bearing surface thus jeopardizing window performance and possibly shortening the window service life.

Empirical testing in the field and lab both demonstrated window performance that exceeded expectations in the respective tests.

The Window 2 geometry ultimately produced defects in the conical window seat that may deem it an unacceptable design for use in submersible service. The Window 3 geometry produced no such defects and may still prove to be a viable design.

Strain and displacement measurements made during the experimental campaign provided less resolution than desired, but were in family with the FEA results and seem to confirm that the FEA constraints are accurate. Additionally, the FEA results generally showed window performance consistent with published results for similar window geometries.

Currently both experimental windows tested for this study (Windows 2 and 3) are being tested to failure under hydrostatic loading to determine the STCP of the windows.

If the study continues, future work will concentrate on defining the maximum allowable hole size for the honeycomb, improving experimental measurement, increasing sample size to confirm test results, and defining reliability performance.

At the end of the day the goal is to increase viewing area for any person fortunate enough to make a trip to the deep ocean.

VIII. References

[1]Busby, R. Frank, 1976. Manned Submersibles. Arlington, Va: Office of the Oceanographer of the Navy. p. 255-256.

[2]Busby, R. Frank, 1976. Manned Submersibles. Arlington, Va: Office of the Oceanographer of the Navy. p. 255.

[3]Stachiw, J.D., Sletten, R. 1976. Spherical Shell Sector Acrylic Plastic Windows with 12,000 ft Operational Depth for Submersible Alvin. Journal of Engineering for Industry. 98(2): 523-536.

[4]Yavari, Yahid, Iraj, Rajabi, Daneshvar, Farshad, Kadivar, Mohammad Hassan. 2009. On the stress distribution around the hole in mechanically fastened joints. Mechanics Research Communications. 36(3): 373-380.

[5]Dupen, Barry. 2014. Applied Strength of Materials for Engineering Technology. Manufacturing Engineering Technology Faculty Publications. Indiana University-Purdue University. 8-2014: 152 total pages.

[6]Jones, Robert M. 2013. Mechanics of Composite Materials. New York, NY: Taylor and Francis Group. p. 400-424.

[7] Hibbler, R.C. 2011. Mechanics of Materials. New York, New York: Prentice Hall. p. 251-271.

[8] Mori, Ken-ichiro. 2001. Simulation of Material Processing. Lisse, Netherlands: Swets & Zeitlinger B.V. p. 891.

[9]Stachiw, J.D. 2003. Handbook of Acrylics for Submersibles, Hyperbaric Chambers, and Aquaria. Flagstaff, Az: Best Publishing Company. p. 314-316, 337-339.

[10]Stachiw, J.D. 2003. Handbook of Acrylics for Submersibles, Hyperbaric Chambers, and Aquaria. Flagstaff, Az: Best Publishing Company. p. 297-350.

[11]Stachiw, J.D. 2003. Handbook of Acrylics for Submersibles, Hyperbaric Chambers, and Aquaria. Flagstaff, Az: Best Publishing Company. p. 289-290.

[12]Imai, Y, Ward, I.M. 1985. A Study in Craze Formation in the Fatigue Fracture of Polymethlmethacrylate. Journal of Materials Science. 20: 3842-3852.

[13] Clark, T.R., Hertzberg, R.W., Mohammadi, N. 1993. Fatigue Mechanism in PMMA at Threshold: Effects of Molecular Weight and Mean Stress. Journal of Material Science. 28: 5161-5168.

[14]Kemper, B. Advances in Acrylics and Expansion of PVHO Window Cyclic Life. In: Proceedings of Underwater Intervention 2016. Underwater Intervention Conference 2016. New Orleans, La: Marine Technology Society.

MTS Manned UW Vehicles 2017 Global Industry Overview

William Kohnen, Chair Manned Underwater Vehicles Cmtee
Marine Technology Society
Jan 24, 2017

Competitive innovation, engineering advancements and leadership are required elements for any long-term business growth, and are essential capabilities for the success of any ocean exploration initiative. Today the ability to design, manufacture and operate human occupied vessels to meet the challenges of extreme ocean environments is driven by a unique group of innovators. Those innovators are part of the elite group that is the Marine Technology Society's Manned Underwater Vehicles group (MUV).

Over the past decades, advancements by the MUV industry have had an impact on a number of programs including space, defense, airborne and other subsea initiatives. These new engineering designs have been implemented to leverage the physics of underwater propulsion, new material fabrication (in the areas of titanium and carbon fiber), and energy storage. There have been advances and testing of new optical cameras for subsea filming and data capture, as well as underwater navigation systems. Today manned vessels work alongside Remotely Operated Vehicles (ROVs), and Autonomous Underwater Vehicles (AUVs) to complete successful ocean mission objectives.

The companies listed below are the leading manufacturers and/or operators who work in this industry and impact the fields of exploration, engineering, design and technological innovations for the 2016-2017.

2017 Manned Overview; alphabetically by Company

Aquatica Submarines, Canada

Sub: *Stingray 500*

Founded: 2013

Vancouver Canada based Aquatica is the newest company to enter the market as an operator and manufacturer. Aquatica's 3-person Stingray 500 is the first in a series of models that has completed a "live drop" to 690 feet for its sea trial with the DNVGL classification society in September 2016. Stingray 500 has a small footprint and weighs 8300 lbs., making it versatile for a wide range of crane types over 4.5 tons. Aquatica submersibles feature a top hatch ingress/egress with a high freeboard and good surface stability to avoid the need for a "man-rated" crane. Two weeks after the sea trial testing, the company completed its first commercial expedition off the coast of Vancouver. The expedition was contracted by the Artificial Reef Society of British Colombia to collect data and imagery while observing artificial reef biomass of the sunken vessel HMCS Annapolis. Aquatica recently completed the design of the Stingray X a 3300-foot submersible for the commercial market.

China Ship Scientific Research Center (CSSRC), P.R. China
Sub: JIAOLONG, Huan Dao Jiao Long

Founded: 1951

CSSRC is China's largest ship and ocean engineering research institute possessing more than 500 research engineers and 22 large-scaled test facilities, mainly engaged in the research of ship design and high performance, underwater engineering, vibration and noise reduction, underwater structure, marine and offshore structures as well as information technology. CSSRC designed and launched the 7000m rated JIAOLONG submersible in 2012 to become the deepest operating manned submersible in the world today.

CSSRC is developing a number of new vehicles including a 4500m rated research MUV for the China Academy of Sciences, with a launch date of 2017/2018. In 2016 the shipyard delivered its first in-country built acrylic-hulled tourism submersibles, the Huan Dao Jiao Long 1 and 2. CSSRC designed, manufactured and tested a set of twin ten passenger vehicles for operation on Hainan Island. The island is a renowned vacation destination in China due to its warm waters and clement weather. Since 1993, the island operated the Meireniyu, a Mobimar Mark III, 44-passenger tourism submersible built in Finland. Due to the long term success of the tourism operation, the two Huan Dao Jiao Long submersibles were commissioned and launched in early 2016. They submersibles are classed by the China Class Society (CCS).

DeepFlight, USA
Sub: DeepFlight Super Falcon, DeepFlight Dragon, DeepFlight Dragon 3S
Founded: 1996

DeepFlight was founded by Graham Hawkes in 1996 to develop innovative underwater technologies to enable exhilarating flight through blue space. DeepFlight submersibles are winged submersibles designed to be "positively buoyant," meaning they naturally float to the surface and use hydrodynamic forces from the wings to push the vehicles underwater when moving. By applying the principles and dynamics of flight, DeepFlight crafts dive using downward lift and the power of its thrust. The submersibles are designed for speed, small displacement and light weight. DeepFlight has developed five generations of personal submersibles, focused on the commercial markets of superyachts, luxury resort and private ownership.

In September 2016 at the Monaco Yacht Show, DeepFlight introduced the DeepFlight Super Falcon 3S, a vehicle design for the tourism market. An extension of the flagship Super Falcon, the 3S model adds a third cockpit so a pilot can take two passengers on underwater flights to depths of 400ft (120m). The new submersible is planned to have Lloyds Register classification and be US Coast Guard certified. DeepFlight also announced a new partnership with Shanghai-based Rainbowfish Ocean Technology called DeepFlight Adventures. The partnership will provide submarine adventures, including resort-based dives, as well as offshore expeditions starting in 2017. The first resort locations are planned to be in Hawaii followed by the Maldives. DeepFlight also plan to jointly develop underwater tourism opportunities with Rainbowfish Ocean Technology. The partnership brings together Rainbowfish's 4800 ton research ship MV Zhang Jian and DeepFlight submersible assets to offer resort-based submersible dives, as well as offshore expeditions. In July 2016 DeepFlight completed a successful joint expedition during the MV Zhang Jian's maiden voyage in Papua New Guinea.

GEOMAR Helmholtz Centre for Ocean Research, Germany

Sub: JAGO

Founded: 1989 (Max Planck Institut)

JAGO is a 400 m depth rated 2-person submersible primarily dedicated to exploration and research in marine sciences. The submersible's relatively light weight (3 tons) and its compact size (3 x 2 x 2.5 m) make it easy to operate worldwide and from a wide variety of support ships that have sufficient crane capacity. It also requires simple logistics for transportation in a single 20-foot sea freight container. JAGO is stationed at the GEOMAR Helmholtz Centre for Ocean Research since 2006 and is presently the only manned research submersible in Germany. The craft was built by its operators in 1989 and is DNV-GL classed. Throughout its career, JAGO has to date made over 1300 dives, diving more than 600 scientists, engineers and observers to the seafloor.

In 2016, JAGO was involved in an expedition with the German research vessel POSEIDON to El Hierro, the youngest and geologically most active of the Canary Islands. A team of scientists from GEOMAR, the University of Las Palmas de Gran Canaria and the Instituto Español de Oceanografía used JAGO to inspect a submarine volcano that erupted offshore at a distance of two kilometers from the southern tip of the island. With JAGO the team documented the present stage of the eruption site, where warm water is still discharged, and sampled hydrothermal fluids, bacteria and other volcanic products.

A second expedition in 2016 took JAGO to the Norwegian Archipelago of Svalbard in the Arctic Ocean. An international team of scientists, led by Senckenberg am Meer and GEOMAR, conducted dives in the Mosselbukta at the very north of Spitsbergen, the largest island of the archipelago, and the Børnøy-Banken south of Spitsbergen to investigate the role of polar carbonate factories mainly composed of coralline red algae and carbonate producing invertebrates.

GlobalSubDIve, USA
Sub: Nemo and Nomad (Triton 1000/2 models)
Founded: 2014

GlobalSubDive provides turnkey underwater exploration solutions with access to advanced technical diving teams in over 54 countries. GlobalSubDive completed three significant missions in 2016 to support global conservation efforts for Project Baseline. The first expedition was with Miami Waterkeepers. The expedition focused on increasing the understanding and awareness of coral reef conditions near Port Everglades in advance of a dredging to expand the port. SCUBA divers collected valuable baseline data from three kilometers of the Florida Reef Tract surrounding the Port Everglades shipping channel. Renowned explorer and environmental advocate, Philippe Cousteau was part of the project and brought international attention to the initiative.

The second expedition teamed with Nekton and a group of leading scientists to execute the first mission of the XL Catlin Deep Ocean Survey in Bermuda. Two submersibles and ten technical divers explored and documented 5 sites at depths between 15 and 300 meters with nearly 100 dives conducted in 24 days. The team produced 138-baseline video transects, and identified nearly 300 species of coral, sponge, and algae. Along with the participation of scientists, journalists and media personalities, a UNESCO representative announced the world's first High Seas World Heritage Site, the Sargasso Sea, from inside the MUV.

GlobalSubDive's third expedition partnered with NOAA's National Marine Sanctuaries and Office of Exploration and Research and the Bureau of Ocean Energy Management to conduct laser and photogrammetry surveys of wrecks that sank in battle off of North Carolina in WWI. Submersibles explored the U-576 and Bluefields both at ~240 meters depth while technical divers explored the wreck of the YP-389 at 90m. Work completed will advance NOAA's goal of expanding the Monitor National Marine Sanctuary and foster ongoing collaborations between NOAA and Project Baseline.

Hawaii University Research Lab (HURL), USA
Sub: Pisces IV & Pisces V
Founded: 1980

The Hawai'i Undersea Research Laboratory (HURL) specializes in providing tools and expertise for scientists to investigate the undersea environment, including submersibles, remotely operated vehicles, and other deep sea technologies. HURL operates the twin Pisces IV and V deep submersibles which are among the longest operating submersibles in the industry. The submersibles are three-person vehicles rated to a depth of 2000m and are ABS classed. The Pisces submersibles have explored all the Hawai'i islands and have traveled across the Pacific Ocean. The university center has had many challenges since NOAA defunded its submersible program in 2014. With the re-certification of its LRT launch and recovery system, HURL was able to significantly reduce its overhead cost for coastal expeditions by not requiring the presence of the R/V KOK. The LRT and the submersibles are both in ABS class at the moment.

2016 looked bright at the beginning of the year with a planned contract with the China Academy of Sciences that was in great need for some deep exploration in the South China sea. With the advent and popularity of JIAOLONG in China, the field of deep ocean research has seen a high level of interest from all research and universities across the country. With the limited access to JIAOLONG, the China Academy of Sciences (CAS) is eager to have access to deep ocean manned vehicles. The China Ship Scientific Research Center is actively building a 4500m submersible for CAS and is expected to be operational in 2018. In this interim, CAS was ready to contract with the PISCES for a 2 month expedition. HURL invested heavily in preparing both submersibles and the KOK of the expedition, which was then canceled a week before sailing. The expedition was caught in political tensions between the US and China over the South China Sea and left HURL in a difficult position. It also left hundreds of researchers in China with a year of preparation and planning hanging, as was expressed in a visit to Shanghai in summer of 2016.

In summer 2016 HURL sailed on an expedition with National Geographics to film corals and seamounts. This was followed by an expedition with Conservation International for a series of 18 dives in 10 days in Loihi with PISCES IV and V and R/V KOK. In late fall 2016, HURL sailed out again for a National Science Foundation expedition funded for Deep Coral Research in the Northwestern Hawaiian Islands. This was for a 30 day expedition using Pisces IV and V and the

KOK for work that had originally been planned to be performed by ROV. The MUV operations proved to be superior and far more effective visiting 5 different seamounts in 12 days of diving, before having to retreat in front of an oncoming typhoon. The success and efficiency of the expedition has ensured that the expedition to the Northwestern Hawaiian Islands will continue into 2017 for deep coral diving research.

ICTINEU Submarins SL, Spain
Sub: ICTINEU3
Founded: 2008

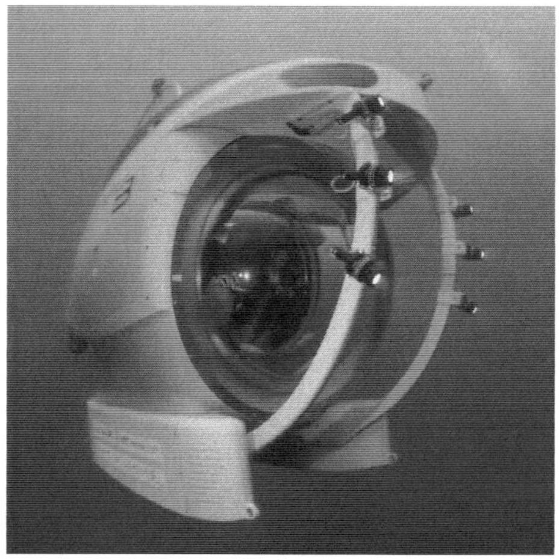

Ictineu Submarins is a manufacturer, operator based in Barcelona, Spain. The ICTINEU3 is the first research submersible designed and built in Spain. The submersible is rated to depths of 1,200 m (3,900 ft), which makes it the ninth deepest submersible. Commissioned in 2013, the submersible carries one pilot and two passengers for dive durations of 10 hours. The submersible's main features include the largest acrylic viewport dome in operation, a highly efficient hydrodynamic design, light weight and a powerful lithium battery module, commercially available today. Ictineu 3 was classed by DNVGL and launched in 2013 after 10 years of development. After completing certification the submersible was registered by French Maritime Affairs for operation in European waters and ICTINEU 3 has completed more than 80 dives at sea between 30 and 1,000 meters depth.

In 2016, ICTINEU3 undertook its first nautical archaeology expedition in the Aegean Sea for a period of 3 weeks. In October, the submersible undertook another archaeological mission in the Mediterranean, 100km north of Barcelona. The mission was lead by the Subaquatic Archaeological Centre of Catalonia, CASC, the Catalan Government agency for underwater archaeology to explore an area that had been poorly explored before in order to find and document new shipwrecks. The center was used to explore with divers limited to 40m depth and 30 minutes dive time. With nearly unlimited dive time at 100-200m, archaeologists discovered three new wrecks from the Roman Imperial era with one expected to be carrying 1,000 to 2,000 amphorae full of "garum" a fish paste that was very popular in the era among the roman court. The lids of the amphorae were still visible and sealed.

Special challenges included navigation between two adjacent ports along the cost, 12 km apart. With rough surface wave and wind conditions, the submersible was used to dive to 45m depth and travel 12 km underwater between the ports. The Battery Bank power provided sufficient power to cover the full distance with a 1 kt head current and arrive at destination with 55% charge capacity remaining. This high level of endurance from the pressure tolerant LiPo batteries allowed the research teams to navigate and explore with great confidence over the entire coastline between the two ports of Sant Feliu de Guíxols, and Palamós. Based on these performance data, the next objective is to perform a 40km transect underwater at 200m depth.

One the technology side, it is typical for the MUV industry to take technology from the ROVs or defense industry (lights,cameras, instrumentation, navigation equipment, etc). In the case if the ICTINEU battery system, it is a rare reversal where a very specific technology developped for a manned submersible is now being transferred to the AUV and offshore industry applications.

ICTINEU SL confirmed it has successfully completed the qualification of its pressure toleramt LiPO battery to 6,000 meters depth. The battery module is designed to be compliant with DNVGL classification rules for construction of submersibles. The ICTINEU LiPo battery pack is a compact pressure compensated battery module offering excellent weight and volume to power ratios, 148 VDC, 10.36 kWh in a 62 liter displacement box containing the BMS, the control circuit, the protective elements and the charge and main contactors. ICTINEU has sold multiple packages to vehicles in Asia and North America for manned and unmanned applications, one of the rare times when technology for MUVs is transferred to the unmanned sector.

JAMSTEC, Japan
Subs: Shinkai 6500

Founded: 1973

Shinkai 6500 is the flagship deep ocean submersible for the Marine Technology and Engineering Center of JAMSTEC (Japan Agency for Marine-Earth Science and Technology). Shinkai 6500 was built in Kobe Japan in 1989, a three person deep ocean research submersible rated to a maximum depth of 6500m. The SHINKAI 6500 is operated from its mothership R/V Yokosuka and has been operated by a pilot, a co-pilot and one observer. In 2016, JAMSTEC has reviewed its operating procedures and is migrating to a concept of operations of one pilot and two observers, as is more common on other deep research submersibles. This involves a reconfiguration of the personnel sphere as well as pilot training changes and new operating procedures.

JAMSTEC initiated a multi-year major upgrade program to SHINKAI 6500 in 2012. This year the overhaul includes a complete dis-assembly of the submersible and two major upgrades. First a change to the personnel sphere layout to accommodate two observers. The second is the installation of an upgraded underwater navigation system. The SHINKAI 6500 operated with a SSBL System and upgraded it to include a DVL sonar. While this improved navigation, the new upgrade includes the integration of an Inertia Navigation System (INS).

For 2017, JAMSTEC plans several test dives in March and April for testing of the new equipment and modifications. This will also include a program of pilot training for single pilot operation of the submersible. The rest of the year is projected to optimize both the submersible equipment and new operating procedures for new expeditions in 2018.

Niggeler-Rebikoff Foundation, Portugal Azores
Subs: LULA 1000

Founded: 1990

LULA1000 is operated by the Rebikoff-Niggeler Foundation (FRN) on the Island of Faial – Azores Archipelago/Portugal. The submersible operates in near open ocean conditions most of the time and was specifically designed for high quality deep sea filming to a depth of 1000 metres. In the fourth year of operation, the LULA1000 has proven to be an excellent and reliable tool for filming and scientific work and continues to be optimized for this task.

In 2016, LULA 1000 performed several Habitat Mappings. Habitat mapping activities to 1000 meters depth were carried out along all the Azores islands´ slopes. Part of the surveys was carried out within existing co-operations, with the Regional Government of the Azores – Directorate for Sea Affairs, and with the University of the Azores.

Dive operations were done along the islands´ slopes of Faial, Pico, São Jorge and Santa Maria. Several new hotspots of marine life and ecologically relevant habitats were discovered and documented with the LULA1000. Several deep sea species could be documented live, in-situ for the first time ever. Located on Faial/Azores, in the central northeast Atlantic, the LULA1000 infrastructure provides fast access to a great number of habitats of ecological importance (e.g. seamounts, deep-sea coral gardens and reefs, sponge aggregations) assuming a strategic position for marine research actions.

Oceangate Inc., USA

Sub: Antipodes, Cyclops 1 , Cyclops 2 in development

Founded: 2009

OceanGate Inc. is a privately held company in Washington that provides subsea manned submersible solutions for industry, research and exploration. The company designs, manufactures and operates submersibles. OceanGate owns and operates the 5-person Antipodes MUV rated for 500 meters depth and Cyclops 1, a three person submersible also rated to 500m. It is also designing and building Cyclops 2, a manned submersible to take a crew of 5 people to depths of 4000 meters. Among several innovative features, its main pressure vessel will be carbon fiber, the first of its size for a manned submersible. The project is scheduled for launch in 2018.

In 2016 OceanGate conducted manned submersible expeditions to the wreck of the Andrea Doria, to the Greater Farallones National Marine Sanctuary, to Desolation Sound in British Columbia, and conducted the first manned submersible dives around Alcatraz Island. Plans for 2017 include expeditions to The Bahamas and to the wreck of the Andrea Doria near Nantucket. Oceangate is also developing Ms.Lars 2 (Mobile Subsea Launch and Recovery System) for launching and recovering manned submersibles and shock sensitive equipment such as environmental sensors and electronic data collection devices.

Rainbowfish Ocean Technology Co., Ltd, P.R. China

Sub: Landers, ARV, Rainbowfish 11000 in design

Founded: 2014

Rainbowfish is a new venture in China leading the research in deep sea technology and their marketing developments. Rainbowfish is the operations and marketing group working with the Hadal Science and Technology Research Center (HAST) of Shanghai Ocean University, which endeavor to challenge the 11000m depths of the Marianna Trench within the next 4 years. The operation is particular as it is a private organization, to the extent possible and is not a state agency. This is quite innovative on its own and the mission objectives are daring.

Rainbowfish privately financed and launched its 4,800-ton mothership MV Zhang Jian in summer 2016 with plans to become an international ocean laboratory, operating a range of manned and unmanned vehicles, conduct scientific expeditions as well as engage in commercial subsea work and make tourism a significant part of its business. In August 2016 Rainbowfish completed part of the sea trials of its MV Zhang Jian mothership in Papua New Guinea which included a joint expedition with DeepFlight.

In 2016, HAST Research Center has continued to test its deep landers, the development of its ARV (Autonomous Remotely operated Vehicle), a form of hybrid ROV and the R&D work in its development of an 11,000m manned submersible. The Full Ocean Depth testing of the ARV is scheduled for late 2016 and completion of its 11,000m submersible by 2020.

SEAmagine Hydrospace Corporation, USA
Sub: *Ocean Pearl, Triumph, Aurora 3/200, Aurora 3/1500, Aurora 3C, Aurora 4/330, Aurora 5/330, Aurora 5/1000, Aurora 6/1000*

Founded: 1995

SEAmagine is a California based company established in 1995, manufacturing manned submersibles with an existing fleet of ten submersibles worldwide having accrued over 12,000 dives. SEAmagine's submersibles are used in scientific, commercial, and superyacht sectors and have featured in numerous National Geographic, BBC, and other film projects. SEAmagine has built a reputation for design excellence, a perfect safety record, and a solid track record in reliability and practicality in both the professional and superyachts markets.

In April 2016, SEAmagine marked another historical milestone with the first manned submersible delivered to South America. SEAmagine delivered an Ocean Pearl, two-person work submersible to the Argentina Coast Guard. The unique project was years in the organization but was finally completed in 2016 to a grand reception in Buenos Aires. The submersible will be used for Law Enforcement both at sea and in its many remote inland lakes. The custom trailer compatible design allows great flexibility to easily transport the vehicles to coast regions as well as deep, remote lakes in the Andes Mountains. Training of coast guard officers is scheduled to continue into 2017.

During the 2016 Monaco Yacht show, SEAmagine launched the AURORA submersible line with the 6 person AURORA-6. The 6 person luxury model offers a very spacious environment for the occupants as well as offering exceptional underwater field of view. The submarine's cabin is arranged in four segments enhancing privacy with the pilot at the front, the owner's lounge right behind, the guests' lounge in the rear, and includes a private emergency restroom in the back hull compartment. Depth ratings range from 150 m to 1500 m. All the company's submarines are classed by the American Bureau of Shipping (ABS) and approved by the Cayman Island Shipping Registry.

Stanley Submarines, Honduras
Sub: IDABEL
Founded: 1998

Founded by Karl Stanley, Stanley Submarines goal is to provide the most cost effective direct access to the deep sea available to the public. Considered one of the longest operating and deepest diving subs the public has had regular access to since 1998. The first sub C-BUG controlled by Buoyancy Underwater Glider retired in 2012 with Idabel launched in 2012 to carry a sub pilot and two passengers to depths of 915 meters or 3,000 feet. The Roatan based company is focused on tourism but the location has provided unique opportunities for scientific research and film and media projects. To date, Karl Stanley logged over 2,000 dives totaling over 5,000 hours.

In 2016 Stanley Submarines conducted over 80 dives. Highlights include filming six gill sharks with both ZDF German TV, and the highest rated show on Animal Planet, River Monsters. Filming was also conducted with the Ukranian travel TV show, Heads and Tails. Dives for science were conducted with staff from the University of Michigan in an ongoing project to study Crinoids and with scientists from the Smithsonian to study deep water fish. One of the farthest tows of the sub to date was done (25 miles roundtrip at 3-4 knots) to investigate the crash site of a plane crash reported to have over 4,000 kilos of cocaine and the co-pilot still aboard. A device nick-named the "sub -sled" is under construction and will be used to streamline the sub and boost the towing speed in an effort to explore areas of interest identified by the multibeam work done by the Schmidt Ocean Institute.

Submergence Group, LLC (USA) & MSUBS Ltd. (UK)
Sub: S201, S301, S351, MASTT
Founded: 2007

Submergence Group and MSubs Ltd provide manned and unmanned submarines and vehicles for defense, research and commercial sectors. The company's capabilities include Pressure Vessel Design and Construction, control and navigation system integration, HP Gas and Hydraulics, Life Support Systems, and GRP tooling. The team's expertise encompasses mechanical, composites, electrical, and electronic and software design. The worldwide network of engineering and material science resources allow both companies to scale quickly based on project delivery cycles.

MSubs recently produced the largest unmanned underwater vehicle in the world weighing in at 63 tons. At 80 feet / 24m long it is used as a submarine tracking target. Last summer, Lockheed Martin and Submergence Group announced a new partnership to build, integrate, test, and deliver up to three dry combat submersibles (DCS) to US Special Operations Command (USSOCOM). The partnership is focused on a $166US million awarded contract that will supply a new class of combat submersibles that will operate at greater depths and with longer endurance. The DSC will support two operators (pilot and navigator) plus up to six swimmers with the ability to lock them out and in.

Triton Submarines, LLC, USA
Subs: TRITON 3300/3, TRITON 1650/3 LP, TRITON 1000/7, TRITON 3300/1 MD, TRITON 1000/2, TRITON 6600/2
Founded: 2008

Triton expanded design and manufacturing operations for 2016 with construction on five new submersibles, including two new models. The Triton 3300/1 MD (minimum displacement) was built for a European client who already owns a Triton 3300/3 and wanted a second submersible to complement their film and documentary capabilities. This one-man vessel can dive to 1000m. The new Triton 3300/3 MKII is the sixth in the product line and is a 3-man vessel. The completed Triton 3300/3 MKII was delivered to the client and used in the Sea of Cortez in July. The vessels feature the new single point lifting system, enabling launch and retrieval in a greater variety of sea states. The first of the new compact Triton 1650/3 LP (low profile) submersibles was launched at Monaco and sent to the Fort Lauderdale Yacht Shows and the Blue Ocean Film Festival in St. Petersburg to highlight its flexibility and capability as a film platform. Triton has also begun construction on a second submersible in this model line and will be using this vessel for charter in 2017.

Tritons submersibles have been used in numerous research and scientific discovery missions around the world, ranging from local Florida surveys to expedition work in the most remote corners of the Pacific. Highlights of these missions include filming the Britannic (sister ship to the Titanic) in the Mediterranean, the first ever manned submersible dives in Antarctica, and the first live footage captured of a giant squid off the coast of Japan.

UBoat Worx, Netherlands
Subs: Super Yacht Sub 3, C-Explorer 3, Cruise Sub, HiPer Sub, C-Researcher 3
Founded: 2005

U-Boat Worx, was founded in 2005 by Bert Houtman is a manufacturer of private submersibles. The company has a wide range of models ranging from 1 to 9 people and operate to depths up to 1,700m. U-Boat Worx submersible are designed, engineered and built to DNV-GL classification.

In 2016, Uboat Worx launched the unique Cruise Sub designed with two large acrylic spheres on either ends. The aerodynamic design of the revolving interior provides passengers the best view possible. The Cruise Sub series can hold 5, 7 or 9 occupants. Both the Cruise Sub 5 and the Cruise Sub 7 submersibles are rated to a depth of 1,140m. The large passenger capacity is specially designed for tourism. For deeper applications U-Boat Worx developed the C-Researcher 3, designed more for research and which offers more equipment options with an array of tools and measurement equipment.

Undersea Hunter, USA
Subs: DEEPSEE (SEAmagine Triumph)
Founded: 1990

The Undersea Hunter Group is based in Costa Rica since 1990 specializing in diving expeditions to remote Cocos Island, located 300 nautical miles off the Pacific coast. The company oprates two expedition vessels, the Sea Hunter for dive expeditions and the ARGO, the mothership for the DEEPSEE submersible. DEEPSEE is an ABS Classed, three-person SEAmagine Triumph submersible rated to a depth of 1500ft. Launched in 2006, DEEPSEE aboard ARGO celebrated it's 10th anniversary, diving year-round in Cocos Island along with several filming expeditions with National Geographics. Having a full global reach the team has easy access to the eastern Pacific region, including remote locations such as the Colombian Island of Malpelo, Galapagos, and the Mexican coasts of Baja California. Along with adventure diving DEEPSEE supports major underwater film projects, to dive producers and film production teams to remote and isolated ocean locations.

In 2016, DEEPSEE continued its collaboration with the National Geographics ""Pristine Seas" expedition. The first expedition was to Clipperton Atoll, a minor French Territory in the Eastern Pacific, 1,581 miles west of Costa Rica. The Nat Geo expedition, led by Dr. Enric Sala, included 10 days of diving on Clipperton Atoll with 18 dives and the collection of sands, corals and hours of high quality video. The expedition continued to the Mexican coast to re-visit Mexico's Revillagigedo Archipelgo, which the DEEPSEE team had first explored in 2014 for a Mexican television production called "Por La Planeta". The Pristine Seas team objective was to document and evaluate the biomass and diversity of fish, sharks and corals of the Socorro Island, San Benedicto Island and the Roca Partida outcropping. The goal was to gather documentation in support of inspiring more government protection for these remote islands. These islands are heavily visited by large numbers of diving liveaboards every year. Of special note, the entire eastern Pacific was under the influence of El Nino and the warmer waters changed the usual migration routes of the humpback whales which normally migrate by the islands from January to April. None were seen during the whole expedition. The 16 day expedition recorded 30 dives with over 100 hours of surveys and dozens of samples for the Nat Geo team. This concluded the 5th expedition with the Nat Geo Pristine Seas project.

Woods Hole Oceanographic Institution, USA

Subs: ALVIN

Founded: 1964

The Woods Hole Oceanographic Institution's Deep Submergence Operations Group operates the ALVIN submersible. Launched in 1964, ALVIN has seen several overhauls. Since 2012, the latest upgrade was completed which included: a completely new personnel sphere rated to 6500m with larger viewports, a new digital command and control system with advanced imaging and data recording systems, an improved maneuvering and control system, an improved cabin interior and better manipulator mounts for better sampling capability. While the personnel sphere is rated and tested to 6500m, the submersible maintains its original 4500m depth rating until the next overhaul phase upgrades the rest of the vehicle. The upgrade was funded through the National Science Foundation, it is certified by the US Navy and the vehicle is owned by the Office of Naval Research.

Since the completion of its first NAVSEA certification to 3800m in 2014, ALVIN was officially certified to 4500m in Jan 2015. To date, ALVIN has completed over 210 dives since its overhaul, with 16 operational cruises. WHOI future plans include the completion of the second phase of the overhaul to bring ALVIN to its 6500m depth rating in the next 4 years. The overhaul is planned for 2019/2020.

Expanding Underwater Technology Applications in STEM K-12 Education Outreach

Jan Rice, Angela Leimkuhler Moran, Sarah S. Durkin, Jennifer A. da Rosa, Rachel Fees

USNA STEM Center for Education and Outreach

Underwater Intervention '17
New Orleans, LA

1

Presentation Outline

- Background and Motivation
- USNA STEM Center
- Underwater Technology STEM Modules
  - Free Glide
  - Pressure Sensors
  - Spuds in Subs
  - Additional modules
- Educator Feedback
- Personal Impacts
- Discussion

All photos courtesy of and/or property of USNA STEM, unless otherwise noted.

2

Background and Motivation

- Every year national leaders reiterate increased need for youth to
  - be equipped with skills to solve complex problems
  - gather evidence
  - think critically
- Project-based learning and hands-on science and engineering
  - addresses real-world challenges
  - encourages creative problem solving
- Ocean engineering, underwater robotics and exploration technologies are an ideal platform for outreach because they are
  - scalable to locally available resources
  - scalable to a wide variety of grade levels
  - introduce K-12 students to the field of underwater engineering and technology
  - can be applied across multiple STEM fields

<u>Challenge</u>: What additional aspects of underwater engineering and technology can we capitalize on to expand K-12 student education in the classroom?

3

USNA STEM Center for Education and Outreach

- Includes 1 Director, 1 Associate Director, 4 staff members, reservists, interns
- Supported by 40+ Faculty members
- **<u>Provides</u>**:
  - Support to all USNA technical majors, Diversity Office, Admissions Office, Community Relations/Public Affairs, Naval Academy Foundation, Office of Naval Research, Office of Secretary of Defense.
  - Opportunities for midshipmen to assume leadership roles and engage in and enhance their undergraduate studies in STEM fields
  - Local and national level K-12 educator and student training programs

2015-2016 STEM Midshipmen volunteers for Girl's Day

4

USNA STEM Center for Education and Outreach Events

SeaPerch Regional Competition and Showcase

Science and Engineering Festivals

MESA Days

STEM Educator Training (SET) Sail Workshops

Girls – Only STEM Events

Summer STEM Camps

Also, STEM Outreach Trips, Mini-STEM Events, Family STEM Days…

Underwater Technology Module: Free Glide

- **Learning Outcome**:
  - Describe basic principles of lift, drag, gravity and thrust.
  - Describe angle of attack and demonstrate how changes to angle of attack will affect resulting glide.
- **Activity**:
  - Construct an underwater glider from triangles of different size foils. Add weights to the foil to adjust its angle of attack and sink rate.
- **Scalability**:
  - Measure and compare angle of attack to glide distance.
  - Measure and compare added weights to glide distance.
  - Graph data, with either MATLAB or Excel, to correlate results.

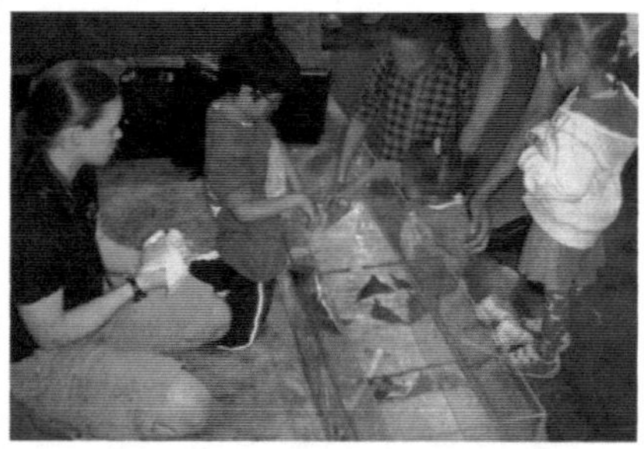

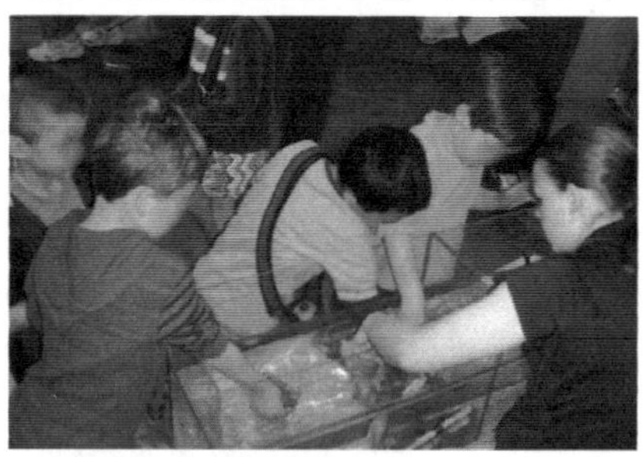

Pictures from 2014 USA Science and Engineering Festival.

6

Underwater Technology Module: Pressure Sensors

- **<u>Learning Outcome</u>:**
  - Describe basic principles of sensors and how they can be used to collect data.
  - Demonstrate relationships between input and output.
- **<u>Activity</u>:**
  - Use cardstock, aluminum foil, an LED, and a battery to create a pressure sensor.
- **<u>Scalability</u>:**
  - Brainstorm list of inputs/outputs and how a sensor can be used to measure desirable outcomes.
  - Discuss how environmental factors can impact sensor readings.

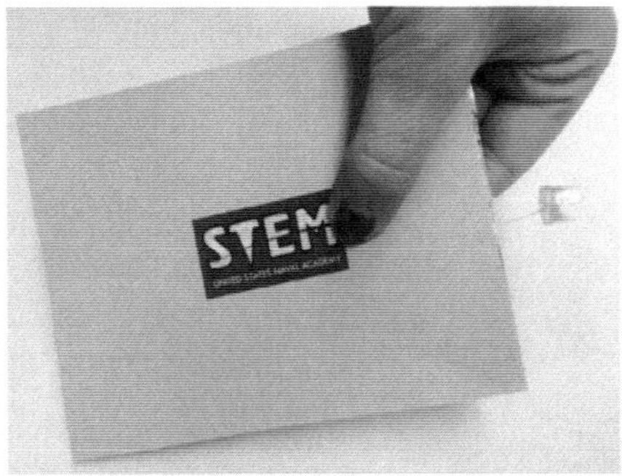

7

Underwater Technology Module: Spuds in Subs

- **<u>Learning Outcome</u>:**
  - Describe the term buoyancy and state factors that affect buoyancy.
- **<u>Activity</u>:**
  - Use a variety of materials to design a submarine hull for a potato so that it neither sinks nor floats, but hovers in the water. Test the design in a tank of water.
- **<u>Scalability</u>:**
  - Calculate/measure and compare weight of displaced water with weight of the potato and designed hull.
  - Graph data, with either MATLAB or Excel, to correlate results.

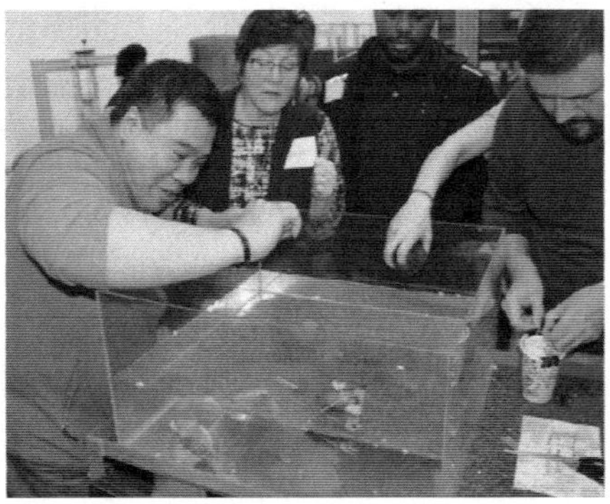

8

USNA STEM Center
Additional Underwater Tech. Modules

Buoyancy
DAS Boats
Flinkers
Coke Float
Dancing Raisins

Electronics & Robotics
SeaPerch
Build Your Own/ Solder
Arduino

Optics
Bending Pencils
Periscope

Acoustics
String Theory
Making Waves
Sound Tubes

Alternate Energy
Power of Wind
Whirly Tube – Hydropower
Sunny Bake Oven

Center of Gravity
Hull Integrity

Environment/ Resources
Marine Oil Spill

Water Properties
Drops on a Penny
Paperclip
Capillary Races

9

K-12 Educator Feedback

General response #1 : Project based learning reinvigorates student's desire to learn and teacher's passion to teach.

General response #2: We need more hands-on, scalable classroom modules.

10

Hands-on PBL
Personal Lessons Learned

What my students have to say:

"Plate tectonics with the marshmallow fluff - it was engaging and I learned."

"Going out on the research vessel was a great experience because it put actual instruments and equipment to the things we learned it class."

My thoughts:

- *PBL has taught me to teach in a way that helps students learn.*
- *PBL provides a sound way to interest students in future studies in engineering.*
- *Without scalable modules, however, PBL can be expensive.*
- *More often than not, most of these costs are absorbed out of pocket – something that many schools and/or teachers can not sustain.*
- *Foundational principles learned in ocean engineering and underwater technologies can easily translate to hands-on, scalable modules.*
- *Countless opportunities can exist for industry partners – but we need to work together to continue to develop and build these opportunities.*

11

Discussion

- Questions?
- Suggestions to improve or expand these examples?
- Other project ideas?

Contact information:
CDR Jan Rice
U.S. Naval Academy
410-293-6524
jrice@usna.edu

12

Experiential Marine Geology and Geophysics Education, White Oak Bayou, University of Houston-Downtown

Alex Barnard[1,4,7*], Philip C. Lyons[1], Alistair Coutts[2], Euan Mackay[3], Michael D. Max[4,5], Georges Detiveaux[6], Jeffrey Martz[1], Eszter Trufan[1], Ajay Singh[1], Jonathan E. Snow[7], William W. Sager[7], Julia S. Wellner[7], Ricky Muñoz[1], Daniel B. Estes[8]

Abstract—**Advanced marine technology is an essential tool used for delivering answers to science-driven questions in underwater environments. Using state of the art technology outdoors encourages intrinsic motivation and provides major opportunities not possible in a laboratory. The University of Houston-Downtown is uniquely situated on the banks of the Buffallo and White Oak Bayous, in closer proximity to an aquatic environment than many world-class marine institutions. Ordinarily, marine research and education focusses on the nearest ocean, lagoon or accessible sea area. We are able to conduct hands-on marine education in a safe shore-based location that is on campus, allowing more of the total time allocated to this activity to be spent learning and operating the equipment rather than traveling. Preliminary underwater investigations of a region of White Oak Bayou that is easily accessible on the campus waterfront were completed in February of 2016 in order to create interest in exploration of natural, and human-made, structures in the local bayou. A Predator (Seatronics Inc., an Acteon company) remotely operated vehicle (ROV) mounted with high-resolution Teledyne Blueview Technologies M900-130 900 kHz 2D multibeam sonar, and optical-video systems was used as a survey vehicle. Where visibility was too low to use optical imaging details of the bayou floor were recorded by the multibeam system. The acquired sonar data clearly images fish, large boulders, and several objects that could be hazardous to even shallow-draft boats and recreational water users. Moving the ROV closer to the submerged features allowed visual inspection and analysis of the video images. The students flying the ROV demonstrated our ability to map the bayou floor, evaluate its ecosystem, detect potential hazards, and identify features of interest. Our methodology, where advanced technologies are sourced from industry partners, provides a practical approach that enables high impact educational experiences for students without necessitating the development of long-term programs, thereby leveraging both valuable research funds and outreach opportunities. A central theme of this work is research experiences for students, who would learn workflows, the scale of operations, and challenges involved in exploratory research.**

Index Terms – **Marine technology, Underwater equipment, STEM, Training.**

[1] Department of Natural Sciences, 1 Main Street Room 813-North, University of Houston Downtown, Houston, Texas, 77002, USA
[2] Seatronics Ltd., Denmore Industrial Estate, Bridge of Don, Aberdeen, AB23 8JW, UK
[3] Seatronics Inc., an Acteon company, 1319 West Sam Houston Pkwy North., Suite 150, Houston, Texas, United States of America
[4] Hydrate Energy International, Kenner, Louisiana, 70065, USA
[5] Department of Geology, University College Dublin, Belfield, Dublin 4, Ireland
[6] Center for Teaching and Learning Excellence, University of Houston Downtown, Houston, Texas, 77002, USA
[7] Department of Earth and Atmospheric Sciences, Science & Research Building 1, 3507 Cullen Blvd, Rm. 312, Houston, Texas 77204-5007, USA
[8] Baytown Police Department, 3200 N. Main Street, Baytown, Texas, United States of America
Corresponding author (AB): E-mail: abarnard2@uh.edu

I. Introduction

Geoscience education often lags frontline research; to overcome this challenge we conducted a survey that used advanced marine technology launched from our on-campus dock. The outdoor environment provides opportunities for discussions in an informal and familiar environment that lends itself to the use of a common language and enables students to interact within the group to construct a more holistic understanding of research that would otherwise be inaccessible [1, 2]. Furthermore, research-based instructional strategies are widely applicable and are part of successful geoscience workforce recruitment [3].

Our novel approach is part of a university-wide initiative to implement high-impact educational practices in student education [e.g., 4] that is being spearheaded by the Center for Teaching and Learning Excellence and Scholars Academy at the University of Houston - Downtown (UHD). Implementation of High Impact Practices (HIPS) involves creation of active systematic techniques that enable student engagement and successful learning [4]. Our work addresses four main HIPS practices, these are: 1) experiences for students, 2) research, and 3) common intellectual experiences, and 4) collaborative assignments. Inclusion of experiential learning, where students learn from direct experience [5, 6], in the existing curriculum is a fundamental goal.

The potential of study areas close to university campuses that could be used for teaching is often overlooked. Ordinarily, when setting a marine teaching or research project, first thoughts for a suitable location may go to the nearest ocean, lagoon, or accessible sea area. The unique location of UHD means that we are able to conduct hands-on teaching in an aquatic environment on campus. From the dockside offshore work can be taught in a shore-based environment thereby maximizing the focus on the training. As time is arguably the most precious commodity, this allows more of the total time allocated to this activity to be spent learning and operating the equipment rather than traveling. By working on campus we can also mitigate risk for the participants and also liability for the university.

Training of industry field crews that undertake survey work was traditionally carried out in house by companies, but recently expectations of graduates require higher levels of training and ready-to-go skills. The reduced capacity for training in industry and increased demand for skilled personnel have resulted in several industry-relevant training courses being offered through professional organizations and at university field camps [7]. Our approach, at the undergraduate level, is to provide hands-on experience with advanced marine technology that encourages intrinsic motivation without requiring a university to have a separate field program. This collaborative and experiential learning approach promotes student performance and critical thinking and is part of creating a meaningful way to increase student engagement in education [8-10].

II. Study Area

The UHD dockside location includes waterfront campus along the confluence of Buffalo and White Oak bayous, providing an exceptional natural laboratory for aquatic science. The word 'bayou' is derived from the native Choctaw word "bayuk", which means creek and the bayou in this region has a rich history but this reach has been overlooked as a potential outdoor classroom or research space due to its present form as a major channel in the city's drainage system.

In 1837, the first steamboat to visit Houston, Laura, took 3 days to make the trip from Harrisburg and could not find the city of Houston because of overhanging trees and log jams on the bayou. On June 10, 1841, the Port of Houston was established at the confluence of White Oak and Buffalo Bayou. The port moved 6 miles downstream when the Houston Ship Channel opened in 1914. It is likely that since the Port of Houston moved location this reach of the bayou, that was once suitable for marine trade, has not been dredged.

The reach of White Oak Bayou adjacent to UHD is 20 – 30 meters wide and prior to the survey no water depth information was available. The bayou generally has a high suspended, and dissolved sediment concentration, and algae growth, that reduce visibility in its waters and obscure the bayou floor. In addition, because of the brown color of the suspended sediments the bayou is often thought of as being dirty. In reality, the bayou contains and sustains a functional ecosystem that is effectively invisible to visual inspection. Because of the buildup of the city, this once natural channel is infrequently overwhelmed by surface runoff and storm discharges that increase point source pollution, and at these times the water quality in the bayou may be lower than usual [11-13].

Built on ideas proposed by Comey [14], Houston's recently invigorated Bayou Greenway Initiative has established a $480 million project to create bayou parks that connect the 10 largest bayous in Houston (www.bayougreenways.org). This is a major environmental project aimed at benefiting the city and its residents [15]. Mayor Annise Parker, and Rich and Nancy Kinder (Kinder Foundation) opened Buffalo Bayou Park on October 3rd, 2015, a park that reaches from UHD to the west of downtown, with further expansion into east Houston planned.

Studying the bayous creates a link between the classroom and outdoor activities that could form part of students everyday lives. The application of marine technology serves a valuable societal, as well as training, purpose. Providing a connection to work that is locally important generates motivation and nurtures students who may come from afar.

A. *Materials and Implementation*

Based on our rationale for field experience, and after gathering students interested in the subject matter, we approached the university to implement safe working operations. Following several safety meetings, we organized an event at which students would use a Remotely Operated Vehicle (ROV) to investigate what lies beneath the murky bayou waters. To prepare for the event we conducted a visual inspection of several sites around UHD to determine the best location for student involvement in bayou research. Several locations were possible, but the dockside adjacent to the Willow Street Pump Station was chosen because of its ease of access for students, equipment, and power. We operated a small portable electricity generator, required to power the ROV and instruments, in a nearby car park (Fig. 1). Prior to launching the ROV we did not have measurements of the bayou depth, but from manual soundings we estimated it to be in the order of one to two meters deep. Similarly, we were aware that there were likely several hazards on the bayou floor. A first pass of the ROV over the survey area at the surface was decided upon so as to avoid and identify any potential hazards at depth.

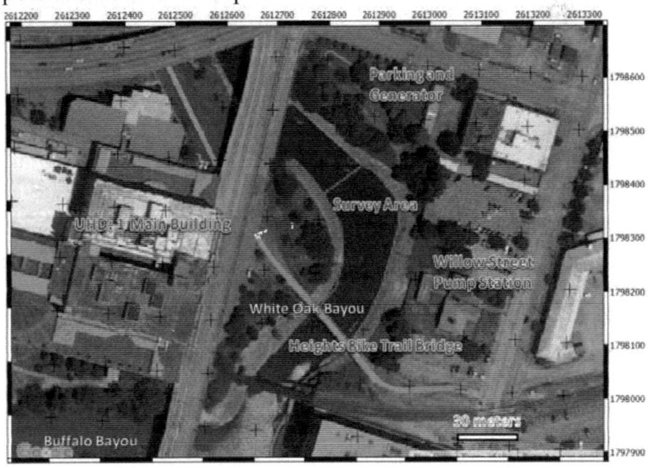

Fig. 1. Location map of the field study area on White Oak Bayou on the University of Houston-Downtown campus. The reach of bayou surveyed is outlined in red and the places used as part of the training are labelled. Image created using open source QGIS (www.qgis.org), NAD83 TX (N) ft. (EPSG: 3668), scale in meters added for reference purposes.

B. Equipment

Equipment is expensive for universities to obtain, maintain and calibrate as it may be used infrequently and is therefore prone to both calibration errors and damage. Equipment is often better sourced from rental companies or industry partners who in turn gain exposure though both the universities students and university. Our methodology, where materials are sourced from industry partners instead of being purchased and left on a shelf for most of the year, provides a widely applicable and practical approach that would allow others in the geoscience education community without long-term programs to provide high impact educational experiences for students, and leverage valuable research funds. Access to marine technology is a limiting factor for most developing marine programs so it is research partnerships and collaboration that stimulate growth [16]. In this case, we collaborated with Seatronics Inc., an Acteon company, to conduct an ROV demonstration on the UHD campus dockside. We used a Predator II (Seatronics Inc., an Acteon company, Houston, Texas) 300 meter rated inspection class ROV that has a maximum speed of ~5 – 6 km/hr. The weight of the ROV is ~80 kg in air. Impellors on the ROV use brushless electric motors from a topside power generator sent to the ROV through the tether. The ROV was mounted with a high resolution wide angle color video camera and a Teledyne Blueview Technologies M900-130 900 kHz 2D multibeam sonar. The ROV also houses several other sensors such as pitch and roll, and a calibrated pressure sensor used to determine water depth.

C. Activities

The event ran from 10:00 – 16:00 in fair weather and drew a crowd; faculty from UHD, University of Houston, University of Houston Clear Lake and ~36 students attended. About ~15 students were present at any one time. The invited guest speaker, Louis F. Aulbach from Texas Historical Commission and the Houston Archaeological Society provided a detailed history of the bayou [e.g., 17]. Two undergraduate students actively participated in helping set up the "topside" equipment, tethering the ROV, and checking the equipment which took about 2 hours, e.g., Fig. 2. This was followed by two hours of hands-on experience with the ROV in the water during which time ~15 students took turns flying the ROV (Fig. 3). Students found they could easily move and maintain position of the ROV even in areas of the channel with relatively strong currents. The ROV technology and its sensors were then presented to students by the visiting industry experts and its capabilities were demonstrated (30 minutes), followed by presentations on the archaeology and history of the bayou, and an introduction to marine geology and geophysics. We allocated as much time as possible for students to get hands-on experience with the ROV, with a professional pilot standing by, throughout the day to focus on student experience and engagement.

D. Exploration

Students were introduced to the ROV, its controls, and instrumentation on the dockside. Each of the ~36 students had opportunities to control the vehicle as we explored the bayou. The first exploratory run was conducted from the surface to gauge the depth and bottom conditions. After determining we could safely conduct a more detailed survey, the students flew the ROV in parallel flights across the bayou to identify targets for more detailed inspection. Two main geophysical sensors were employed: imaging sonar and optical imaging. Combined, these systems provided information on a broad swath of the bayou floor and its banks.

We were able to acoustically image objects ~10 meters in front of the ROV. The sonar could return images of regions underwater where the suspended sediment would not permit optical imaging. Moreover, the video systems provided ease of recognition for targets once the vehicle was maneuvered into a position close enough (~1 meter). As well as controlling the ROV, students were able to manipulate the sensors to improve the images viewed and recorded.

Fig. 2. Launching the Predator II ROV into White Oak Bayou on the UHD campus.

III. Interpretation of Bayou Exploration Data

Exploration of the bayou identified several features that were visible above the water surface and also several interesting features that were not known prior to the survey (e.g., Fig. 4, C, D, & E). There was a great amount of collaborative sharing of ideas and team problem solving to maneuver the ROV into positions to answer student questions about features. We found that the depth of the bayou varied from ~0.5 meters in the northern reaches to 2 meters beneath the Heights Bike Trail Bridge in the south. To student delight, several fish (~10 inches) were observed in the deeper parts of the bayou. We were also able to identify several carts, cables and large boulders in the stream channel that could be potential hazards for recreational water users and small boats.

Fig. 3. Students enjoying working dockside on the ROV topsides.

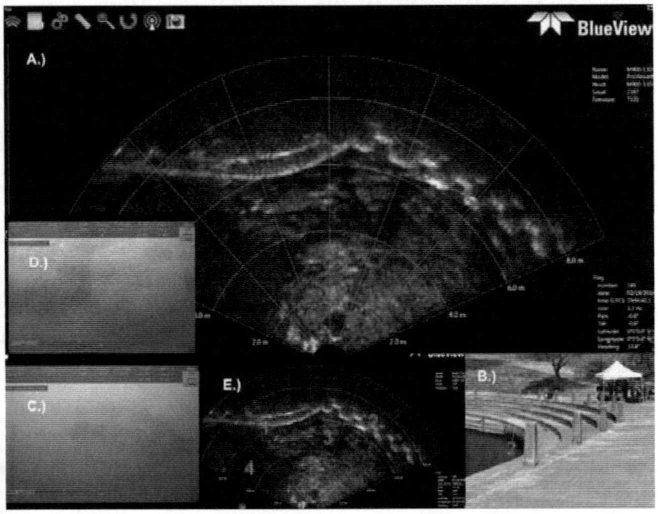

Fig. 4. A) View from multibeam imaging sonar of the bayou floor. In the sonar imagery bright regions represent strong acoustic reflections from either hard materials or objects angled towards the receiver. B) Photograph of the survey region showing: 1) concrete steps, and 2) corrugated metal. C) Largemouth bass (fish). D) Boulders in the stream channel. E) Sonar image with interpretation of significant features: 1) concrete steps, 2) corrugated metal, 3) debris/tires, 4) muddy bayou floor, and 5) fish.

While the ROV was in the bayou waters, the sensors were continuously recording data. This data was synchronously presented on computer monitors and stored on a hard drive. Because we did not set up a location recording GPS base station, the acoustic measurements taken in the bayou were relatively positioned using known locations.

Back in the laboratory, students played back the acoustic imagery using Teledyne Blueview Proviewer v4.3 (Table 1). By matching the time stamps in both datasets we were able to match both the sonar and optical image data. This taught students how to integrate data and the importance of making more robust interpretations with multiple data sets. In some cases, the optical imagery couldn't be used for target recognition due to the low light penetration in the bayou water.

Dataset	Files Acquired	File Size (mb per file)	File type	Software Used
Multibeam	38	~200	.son	Blueview Proviewer v4.3
Video	1	~500	.avi	Windows Movie Maker

Table 1. Information about the files acquired during acquisition and software used for data processing.

IV. Implications

The data acquired during this activity has been used in courses taught at UHD that teach the fundamentals of oceanography and geoscience, and were used in a s tudent poster presentation at student research day [18]. Once students become familiar and interested in marine activities there are opportunities for them to get involved with broader research projects remotely [19] as well as the more traditional industry (internship) and academic (research) routes. Several global oceanographic institutions now broadcast their work through the internet (e.g., NOAA, NAUTILUS LIVE, IODP, NOCS), allowing students direct access to frontline research.

V. Summary

As part of this exercise students were able to get hands-on experience with advanced marine technologies. The day was wrapped up with a debriefing (~1 hour) on the equipment used, the past and present significance of the bayou, and a discussion on data acquisition and interpretation. A student also presented this work at UHD student research day. Because this research was conducted outdoors, it allowed students to engage with marine technology equipment and learn more about the natural environment in a familiar setting that stimulated more questions and discussion than would have been possible in a traditional classroom environment. We identified fish and several objects in the bayou that indicated there is potential in the future for more detailed exploration.

VI. Acknowledgments

Special thanks to Seatronics Inc., an Acteon company, for supplying the Predator II ROV. This work is supported by the Center for Teaching and Learning Excellence (CTLE) competitive High Impact Practices (HIPS) award to AB (2016). We are grateful to Louis F. Aulbach for sharing his expertise on Houston's bayous with us and would also like to thank Buffalo Bayou Partnership (BBP) and the Professional Tour Guide Association of Houston (PTGAH) for up-to-date historical information about the bayou.

References

[1] K. M. Kortz and A. R. Caulkins, "Introductory geology: Is there a common language?," *GSA Today,* vol. 25, 2015.

[2] N. Marcum-Dietrich, L. Marquez, S. E. Gill, and C. Medved, "No teacher left inside: Preparing a new

generation of teachers," *Journal of Geoscience Education,* vol. 59, pp. 1-4, 2011.

[3] N. D. LaDue and H. A. Pacheco, "Critical experiences for field geologists: emergent themes in interest development," *Journal of Geoscience Education,* vol. 61, pp. 428-436, 2013.

[4] G. D. Kuh, "Excerpt from High-Impact Educational Practices: What They Are, Who Has Access to Them, and Why They Matter," *Assoc. of Am. Colleges and Univ., Washington, DC,* 2008.

[5] J. Dewey, *Experience and Education.* New York: Collier Books, 1938.

[6] D. A. Kolb and R. E. Fry, *Toward an applied theory of experiential learning*: MIT Alfred P. Sloan School of Management, 1974.

[7] R. R. Stewart, S. Khan, S. Hall, C. Liner, and J. Wong, "Geophysics field education: Better learning by doing," *The Leading Edge,* vol. 29, pp. 546-550, 2010.

[8] J. Nnodim, "A controlled trial of peer-teaching in practical gross anatomy," *Clinical anatomy,* vol. 10, pp. 112-117, 1997.

[9] M. Breunig, "Turning experiential education and critical pedagogy theory into praxis," *Journal of experiential education,* vol. 28, pp. 106-122, 2005.

[10] M. M. Yacobucci, "Integrating Critical Thinking About Values Into an Introductory Geoscience Course," *Journal of Geoscience Education,* vol. 61, pp. 351-363, 2013.

[11] K. Cantrell, L. D. Leija, J. D. Henderson, J. Vasquez, K. Johnson, and S. W. Slough, "Dissolved metal abundances in bayou waters from Harris County, Texas; a possible monitor of subsidence in the Houston-Galveston area," in *Abstracts with Programs Geological Society of America*, 2007, p. 58.

[12] M. Schmertz, "Houston focuses on the restoration and development of the Buffalo Bayou, a historic waterway," 2003.

[13] T. M. Petersen, M. P. Suarez, H. S. Rifai, P. Jensen, Y.-C. Su, and R. Stein, "Status and trends of fecal indicator bacteria in two urban watersheds," *Water Environment Research,* pp. 2340-2355, 2006.

[14] A. C. Comey, *Houston: Tentative Plans for Its Development; Report to the Houston Park Commission*: Press of Geo. H. Ellis Company, 1913.

[15] J. L. Crompton, "Estimates of the Economic Benefits accruing from an expansion of Houstons Bayou Greenway Network," *Journal of Park and Recreation Administration,* vol. 30, 2012.

[16] M. R. Gilligan, P. G. Verity, C. B. Cook, S. B. Cook, M. G. Booth, and M. E. Frischer, "Building a diverse and innovative ocean workforce through collaboration and partnerships that integrate research and education: HBCUs and Marine Laboratories," *Journal of Geoscience Education,* vol. 55, pp. 531-540, 2007.

[17] L. F. Aulbach, *Buffalo Bayou: An echo of Houston's wilderness beginnings*: CreateSpace Independent Publishing Platform 2011.

[18] R. Muñoz, A. Barnard, A. Coutts, E. Mackay, M. D. Max, G. Detiveaux, *et al.*, "Remotely Operated Vehicle Based Exploration of White Oak Bayou, Main Street Campus, University of Houston-Downtown," in *University of Houston-Downtown Student Research Day,* Houston, Texas, USA, 2015.

[19] A. Pallant, C. McIntyre, and A. L. Stephens, "Transforming Undergraduate Research Opportunities Using Telepresence," *Journal of Geoscience Education,* vol. 64, pp. 138-146, 2016.

Marine Technology Apprentice Program

NORTHWESTERN MICHIGAN COLLEGE
TRAVERSE CITY, MI

Presentation Overview

What are Apprentice programs?

What is the Value Add?

- Employer
- Employee

Northwestern Michigan College

- Training Opportunities
- Facilities & Equipment

Northwestern Michigan College

Apprentice Programs Overview

What is a Registered Apprentice?
Flexible training system that combines job related technical instruction with structured on-the-job learning experiences.

What are the program Requirements?
Registered Apprenticeship programs are a written plan designed to move an apprentice from a entry-level position to full occupational proficiency.

A Registered Apprenticeship program is sponsored by an individual business or an employer association

Credentials
Participants receive an industry issued, nationally recognized credential that certifies occupational proficiency

Northwestern Michigan College

Value Add - Employer

- Customized training that results in highly skilled employees trained to industry/employer specifications
- Enhanced retention
- Emphasis on safety training
- A stable and predictable pipeline for the development of qualified workers
- Recognition of the training program
- A systematic approach to training that ensures that employees are trained and certified to produce at the highest skill levels required for that occupation
- A proven training model that allows employers to set the benchmark and the structure that can determine the Return on Investment in training dollars

Northwestern Michigan College

Value Add - Employee

- Improved skills and competencies that meet the specific needs of the employer
- Incremental wage increases as their skills improve
- On-the-job training and occupation focused education
- Career advancement
- Industry issued, nationally recognized credentials
- Articulation agreements between certain apprenticeship training programs and 2- and 4-year colleges that create opportunities for college credit and future degrees

Northwestern Michigan College

Marine Technology - Apprentice

Overview
Operate, test and maintain unmanned automated marine systems

On-the-job Training
Documented work experiences in targeted competencies

Related Technical Instruction
Identified specific instruction

Northwestern Michigan College

Northwestern Michigan College

Marine Technology - Apprentice

On the Job Training

Content	Hours
Technical (ROV,LAR)	1750
ROV Systems	1250
ROV Piloting	1000
Preventative Maintenance	1000
Projects	350
Fault Finding	200
Documentation / Admin	200
Soft Skills	100
Safety	100
Emergency Response	50
Total	6000

Related Technical Instruction

Content	Hours
Technical •Electronics •Fluid Power •ROV Systems / Ops •Acoustics & Sonar	450
Communications	60
Math	60
Safety	24-40
Total	574

Northwestern Michigan College

Related Technical Instruction

	Fall Semester	Spring Semester	Summer Semester
FIRST YEAR	**Safety – PEC Core Compliance/HAZWOPER 40** Instructional Time: 24-40 Hours **Electrical Studies I** EET 103 Credits: 3 Instructional Time: 60 Hours	**Hydraulics & Fluid Power** MFG 104 Credits: 4 Instructional Time: 75 Hours **English Composition** ENG 111 Credits: 4 Instructional Time: 60 Hours	**Research Technology (ROV & SONAR systems)** WSI 200 Credits: 3 Instructional Time: 60 Hours
SECOND YEAR	**Electrical Studies II** EET 104 Credits: 3 Instructional Time: 60 Hours **Intermediate Algebra** MTH 111 Credits: 4 Instructional Time: 60 Hours	**ROV Systems & Operations** WSI 240 Credits: 4 Instructional Time: 75 Hours **Underwater Acoustics & Sonar** WSI 210 Credits: 3 Instructional Time: 60 Hours	**Marine Diagnostics & Repair** Instructional Time: 40 Hours

Marine Technology - Apprentice
Why Northwestern Michigan College ?
•Facilities
•Instructional Staff
•Vessels & Equipment
•Sensors & Software
•Location
Northwestern Michigan College

Facilities
West Grand Traverse Bay
Dedicated Harbor
Indoor Training Tank
Electronic / Hydraulic Labs
65,000 sq. ft. Building
TRAVERSE CITY
WISCONSIN
CANADA
MILWAUKEE
MICHIGAN
CHICAGO
DETROIT
INDIANA
ILLINOIS
CLEVELAND
OHIO
Traverse City, MI
Northwestern Michigan College

Facilities
West Grand Traverse Bay
Dedicated Harbor
Indoor Training Tank
Electronic / Hydraulic Labs
65,000 sq ft Building
Northwestern Michigan College

Facilities
West Grand Traverse Bay
Dedicated Harbor
Indoor Training Tank
Electronic / Hydraulic Labs
65,000 sq ft Building
Northwestern Michigan College

Facilities
West Grand Traverse Bay
Dedicated Harbor
Indoor Training Tank
Electronic / Hydraulic Labs
65,000 sq ft Building
Northwestern Michigan College
Northwestern Michigan College

Facilities
West Grand Traverse Bay
Dedicated Harbor
Indoor Training Tank
Electronic / Hydraulic Labs
65,000 sq ft Building
Northwestern Michigan College
Northwestern Michigan College

Vessels

Northwestern:

- 56ft research vessel

Hawk Owl

- 19ft vessel

Seafloor Systems Hydrone G1

Northwestern Michigan College

Vessels

Northwestern:

- 56ft research vessel

Hawk Owl

- 19ft vessel

Seafloor Systems Hydrone G1

Northwestern Michigan College

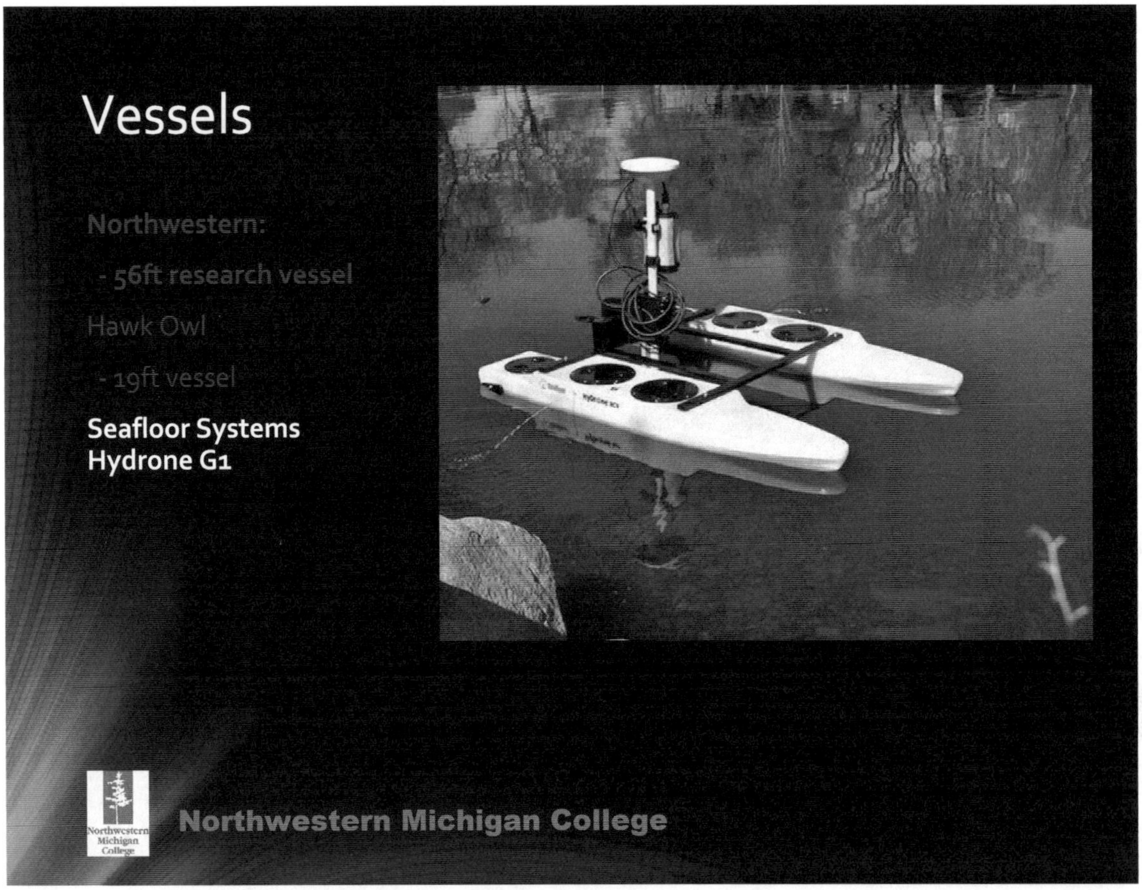
Vessels
Northwestern:
- 56ft research vessel
Hawk Owl
- 19ft vessel
Seafloor Systems
Hydrone G1
Northwestern Michigan College

Equipment
SAAB Seaeye Falcon
ROV
-450M tether integrated
with the M3,
-GreenSea automation &
INS
-5 function manipulator.
Outland 1000 ROV
- 800 ft tether
Northwestern Michigan College

Equipment

SAAB Seaeye Falcon ROV
-450M tether integrated with the M3,
-GreenSea automation & INS
-5 function manipulator.

Outland 1000 ROV
- 800 ft tether

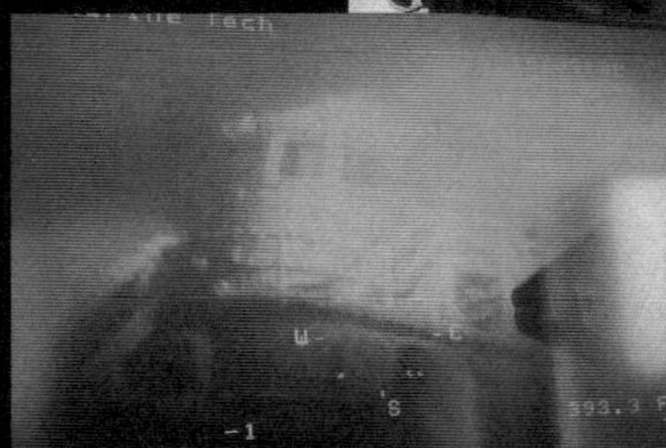

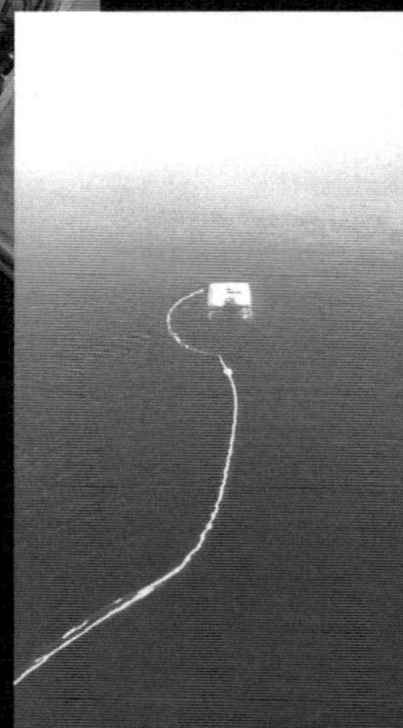

Northwestern Michigan College

Sensors & Software

- Kongsberg M3
- Edgetech 4125 Sidescan dual frequency Sonar
- Kongsberg 1171 Scanning sonar
- Marine Magnetics Explorer Magnetometer
- Velodyne LIDAR
- Linkquest Tracklink USBL
- ORE Trackpoint USBL
- Hypack & Quimera Marine data processing software
- ESRI GIS software

Northwestern Michigan College

Sensors & Software

- Kongsberg M3
- Edgetech 4125 Sidescan dual frequency Sonar
- Kongsberg 1171 Scanning sonar
- Marine Magnetics Explorer Magnetometer
- Velodyne LIDAR
- Linkquest Tracklink USBL
- ORE Trackpoint USBL
- Hypack & Quimera Marine data processing software
- ESRI GIS software

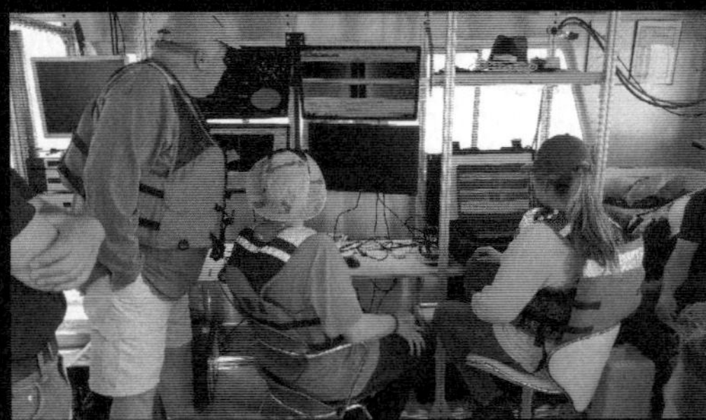

Northwestern Michigan College

Questions

Contact Information:

Dave Jackman
US Department of Labor
Office of Apprenticeship
(517) 377-1746
jackson.dave@dol.gov

Ed Bailey
Technical Division Director
231-995-1215
ebailey@nmc.edu

Marine Technology Camp

Hans Van Sumeren, Northwestern Michigan College
Liesl Hotaling, MTS VP for Education

Goal of the Program

"To promote awareness, understanding, advancement and application of marine technology"

Persistence and a shared vision

MTS focuses on supporting our membership and nurturing the next generation of ocean leaders…

NMC focuses on developing and delivering the next generation of ocean and Great Lakes leaders…

Pilot Plan - 2016

MTS would partner with a marine technology program in the U.S. to offer the first camp experience.

The program was envisioned for undergraduate students, however also open to interested graduate students.

The campers would have several hands-on opportunities to use many technologies during the week long camp.

The first camp would be limited to 12 campers to test the efficacy of the concept.

Pilot Reality

MTS partnered with Northwestern Michigan College in Traverse City, Michigan.

The campers spent time on research vessels, collected data using ROVs, sonar, sensors and buoys and learned about the multiple applications of these technologies.

Eight campers attended:

- 2 University of Florida (female)
- 1 Texas A&M (female)
- 1 University of San Diego (female)
- 1 Stockton University (female)
- 3 U.S. Navy (male)

Monday – BRIEF introduction, Safety, seamanship, MBES, Side Scan Sonar

Tuesday – Scanning Sonar, Shipwrecks and ROV's

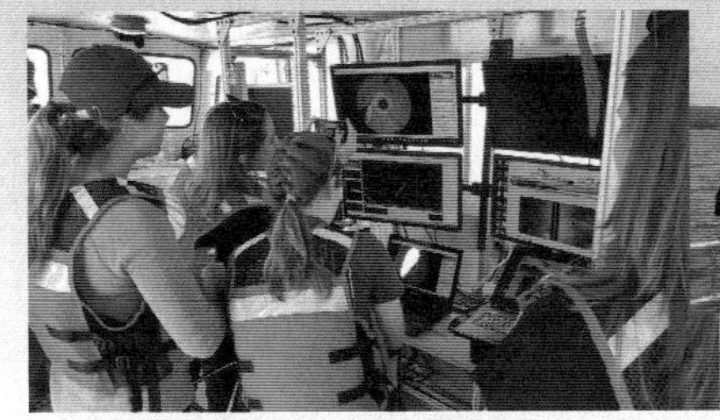

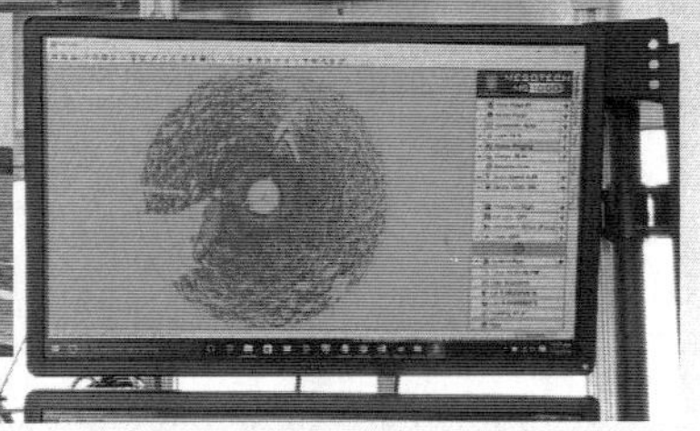

Wednesday – Unmanned Aerial Systems, Marine Technology Labs

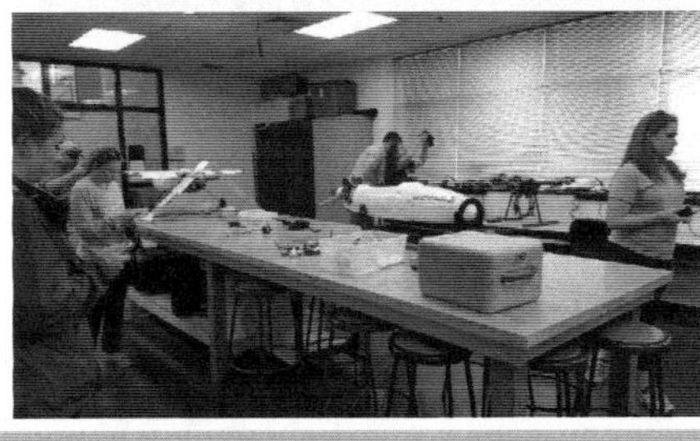

Thursday – ROV's, pipeline surveys, MBES, Harbor Inspections

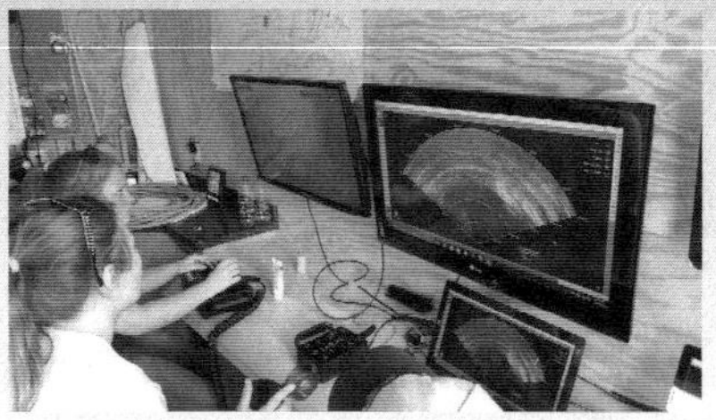

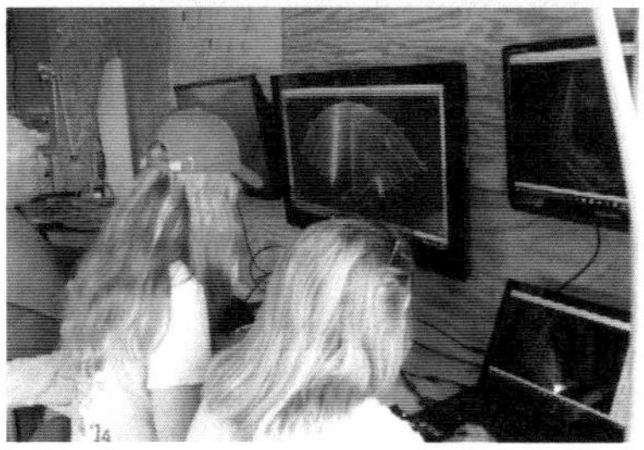

Friday – Multi-ROV operations, Deepwater Shipwreck, Ecosystem Investigations

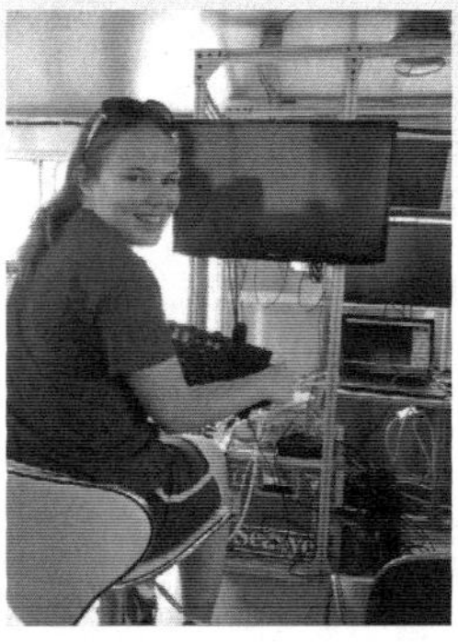

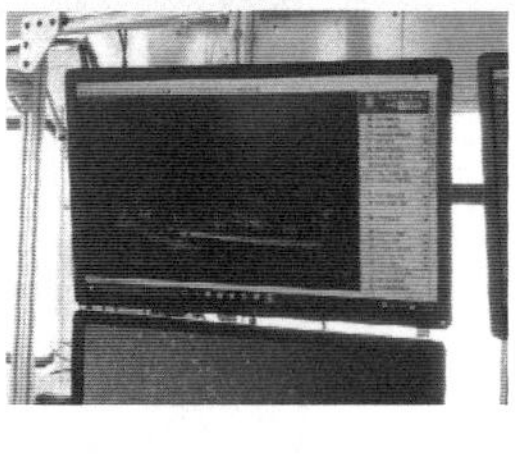

2016 Camp Results

I liked that we spent most of our time out on the lake operating the sonars and ROVs.

I loved being on the water and having the opportunity to experience this technology in the setting that it would actually be used. I really enjoyed using the sonars and scanning the floor. **I had never experienced anything like that before.**

I don't think that any one thing I learned this week was the most valuable. For me, **I came to this camp because I really didn't have a solid understanding of the types of technology used in the marine industry** or the types of work that are encompassed in marine technology.

The most valuable thing about this camp is the wide variety of occupations of the staff and their willingness to explain an aspect of the industry. I loved learning how to use the equipment but **the practical applications that were outlined by the various staff members really helped me gain an understanding of the industry.**

What I liked best was getting hands-on experience using ROVs and different types of sonar, and that the technology and application **was explained to us while we were using it**. I feel like it had **more of an impact learning how to use the equipment WHILE using it** as opposed to getting a how-to lecture and then going out. I also like how we were **shown the basics of using the equipment, but then given the opportunity to play around with it and figure new things out for ourselves.**

In lectures through Texas A&M, we learn about the different technologies (ROVs, sonar, etc) and we're given a short overview of what they are used for in the field; however, **we rarely go over how they work, what they actually do, or how to interpret the outputs they provide**. So at the camp, **it was extremely valuable to learn now only how to use and apply this technology, but also how to interpret the data it presents**. It gave me a **much better understanding of topics we had gone over in my classes from the previous semester.**

I honestly loved everything about this week. I wasn't expecting much, but I was really hoping I just wouldn't feel like the dumbest person there, since my major is more biology based. But I didn't feel dumb at all. **Everyone involved in this program really helped me understand everything we learned.** The professors, mentors, and other students really explained things clearly, and **I feel way more confident about marine technology than I did before I attended this camp.**

Friends for life!

2017 Michigan Plans

24 campers
May 22 – 26, 2017

Future Program Plans

- Expand to Camp 2.0 – students to complete either NMC or Rutgers Camp 1.0
- Add data processing
- Add hydraulics component
- Add electronics component

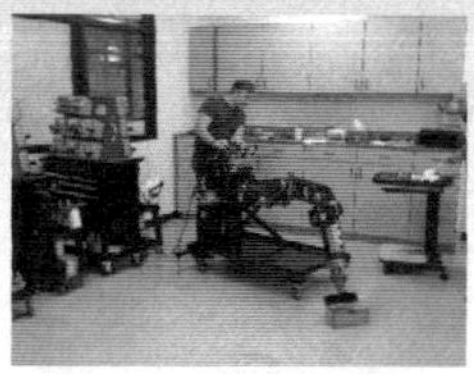

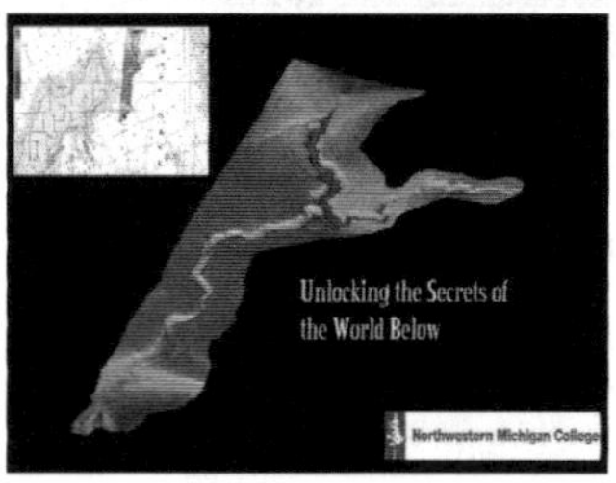

2017 Rutgers Plans

June 12 – 16, 2017

Thank you!

Questions?

Hans Van Sumeren (hvansumeren@nmc.edu), Northwestern Michigan College

Liesl Hotaling (lieslhotaling@yahoo.com), MTS VP for Education

MATE Inspiration for Innovation (MATE II)

The next chapter in the evolution of the MATE international ROV competition

Jill Zande
Associate Director & Competition Coordinator, MATE Center
President, MATE II

MATE II

Background on the Marine Advanced Technology Education (MATE) Center

- The MATE Center was founded in 1997 with funding from the National Science Foundation's (NSF) Advanced Technological Education (ATE) program. It is currently funded in part by the NSF grants, industry donations, and SeaMATE store sales.
- Headquartered at Monterey Peninsula College, **MATE is an international partnership** of 1,000+ educational institutions, marine industries, and professional societies.
- MATE's mission is to use **marine technology** to inspire and challenge students to learn and **creatively** apply science, technology, engineering, and math (**STEM**) to solving real-world problems in a way that strengthens critical thinking, collaboration, **entrepreneurship**, and **innovation**.

NSF

MATE

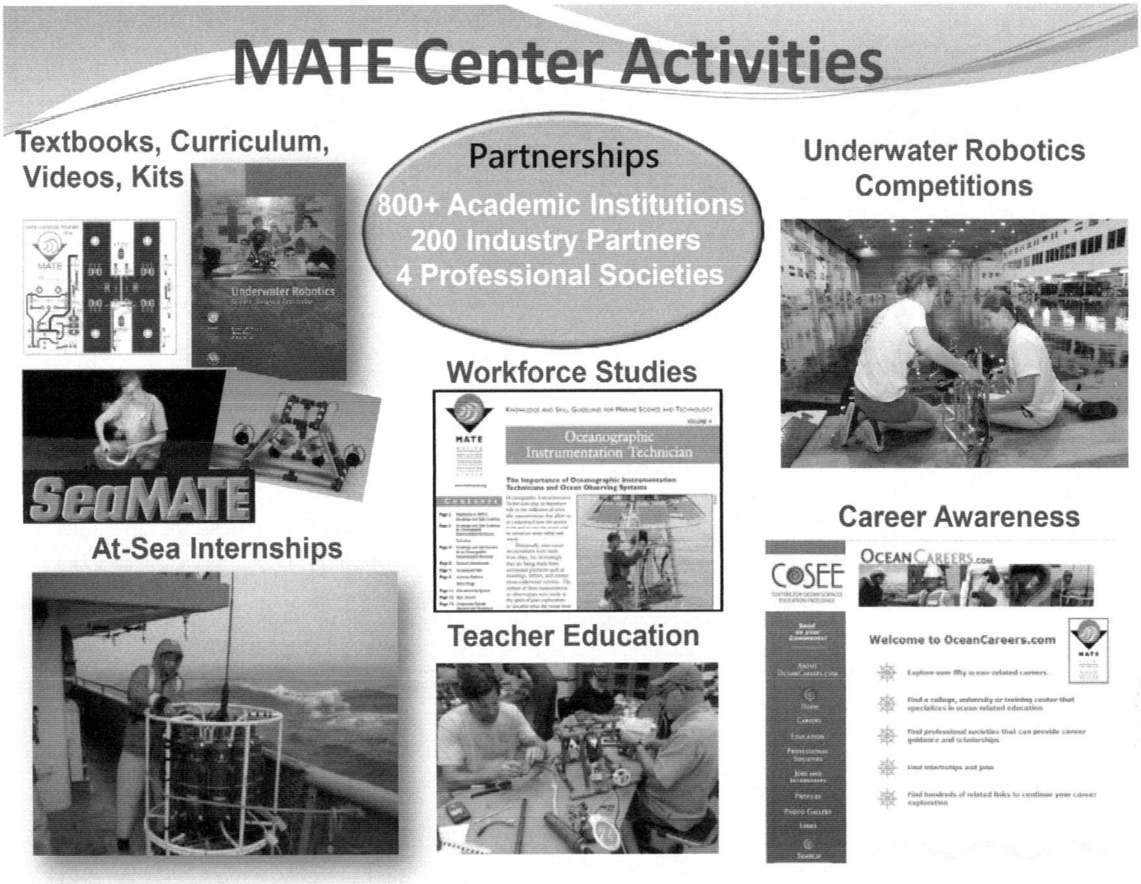

Background on the MATE Competition

Partnering with the MTS ROV Committee since 2000 to…

- **Address shortage of skilled individuals to support the marine industry**
  - Reach more students and get them interested in the industry
  - Help students develop the skills to support marine activities

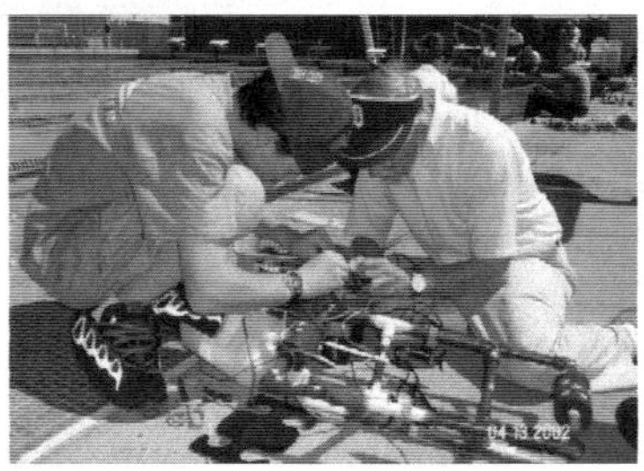

- **Provide a venue for industry to become involved with education**
  - Funds – to support student participation
  - Industry mentors – complement and expand students' and educators' knowledge and expertise

MATE

marine technology SOCIETY

Opportunity runs deep

TO DATE...

- Has involved more than 14,000 students in grades 4-16 who work in teams to tackle missions based on the ocean STEM workplace
- Includes one international competition and a network of **30** (and growing!) regional contests
- Involves 100s of working professionals & organizations
- Includes underwater missions and technical reports, engineering presentations, and poster displays
- Gets students excited about STEM and motivates them to problem-solve, think critically, manage a project, and work together as a team
- Is encouraging students to pursue STEM degrees and jobs in the field
- Has been highlighted in 1 documentary, 1 Hollywood movie, and 1 book

MATE

2017 will mark the 16th international competition!

marine technology SOCIETY
Opportunity runs deep

MATE Regional ROV Competition Network

The MATE Competition Network began in 2001 and currently consists of 30 regional events that take place across the U.S. and around the world.

Use this information to find the regional event near you!

MATE International Regional Competitions:

Bermuda, Canada (Newfoundland & Labrador and Nova Scotia), Egypt, Hong Kong, Indonesia, Scotland, Russia, Turkey

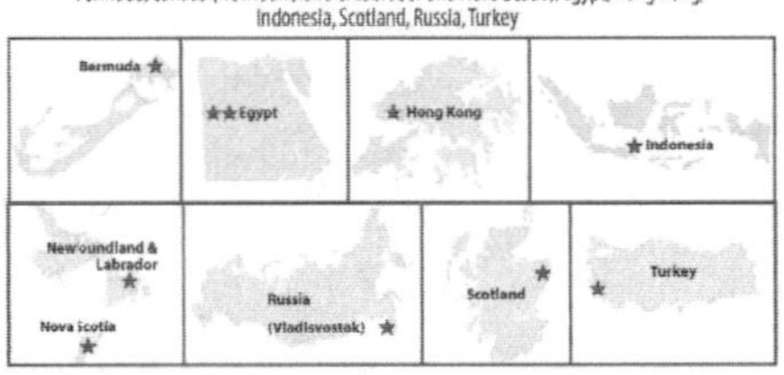

MATE ROV Competition Team Participation 2001-2016

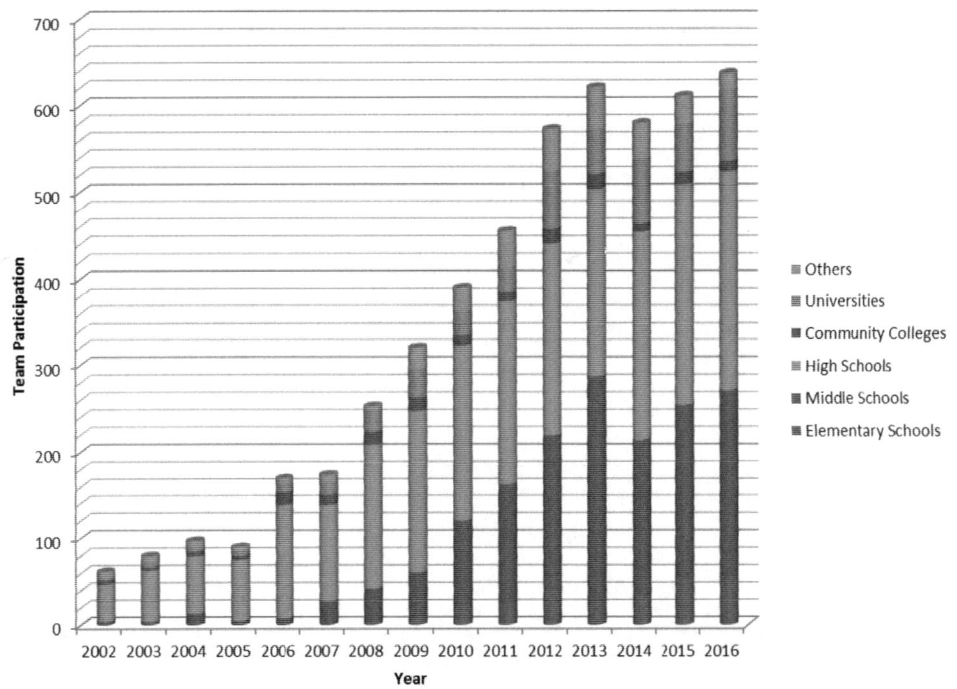

ITEST

- Innovative Technology Experiences for Students and Teachers is an NSF program that focuses on K-12
- The project uses the ROV competition as the focus to develop and support a progression of learning from elementary to middle to high school and beyond
- It was through ITEST that we started creating – then selling – underwater robotics kits

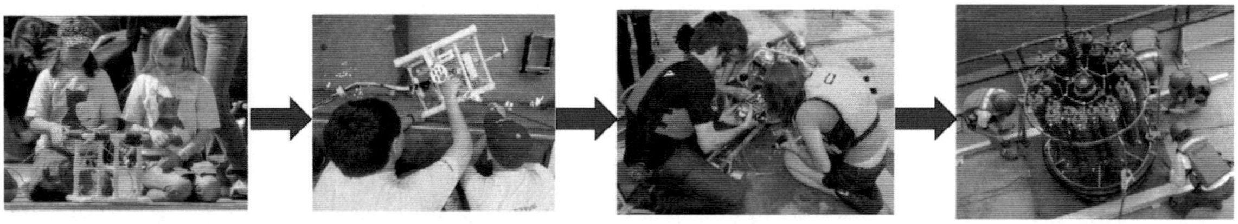

SeaMATE

SeaMATE is a **social enterprise** that provides community college students with workplace experience while creating products and services that promote engineering and technology education.

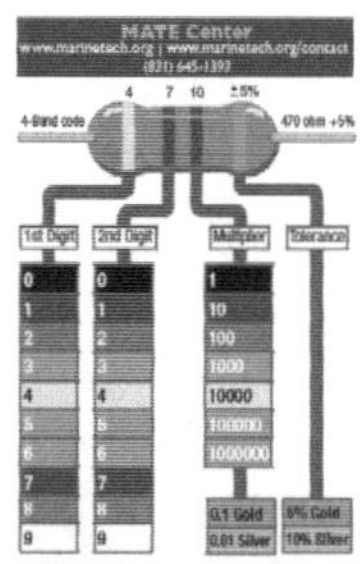

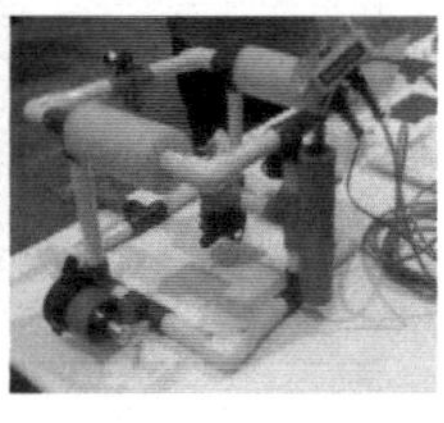

MATE at the crossroads

Challenges can lead to opportunities…

- The competition is growing, but the downturn in the offshore oil and gas industry has meant decreased donations…**need to diversify our donors**
- Encouraged to be entrepreneurial and given some tools (NSF's I-Corps for Learning), but no clear directions on how to make the transition from a grant-funded center to a business…**educating ourselves and paving our own way**

MATE Inspiration for Innovation (MATE II)

- Corporation formed in late October after…
  - Hiring a lawyer to walk us through the process
  - Signing an MOA with our college
- IRS 501(c)3 status pending
- Board of Directors
  - Jill Zande, President
  - Deidre Sullivan, Vice President
  - Debbi Kill, Treasurer
  - Fritz Stahr, Secretary
  - Justin Manley, Member-at-Large

MATE Inspiration for Innovation (MATE II)

- 501(c)3 opens doors previously closed
  - Fundraising efforts
  - Certain states offer funding to schools that participate in robotics competitions organized by 501(c)3s.
- Gives us the chance to:
  - Rebrand – logo, sponsorship package
  - Document processes and procedures and standardize regional practices
  - Explore and adopt new models to manage growth
- Sets us up for "future directions"

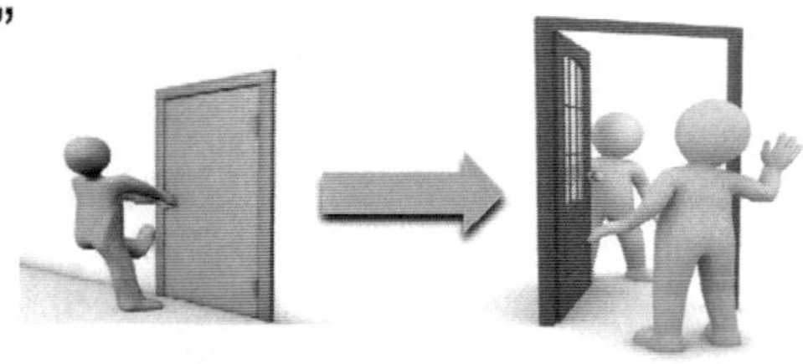

MATE II

Thank you for MAKING A DIFFERENCE by supporting **MATE Inspiration for Innovation (MATE II)**. Your contributions help inspire and challenge students to learn and creatively apply science, technology, art, engineering, and math to solving real-world problems in a way that strengthens critical thinking, collaboration, entrepreneurship, and innovation.

MARINETECH.org

Sponsorship Matrix

MATE II

*Discounted rate available for multi-year commitment

Benefits	Team $1,500	Bronze $2,500	Silver $5,000	Gold $10,000	Platinum $25,000	Diamond $50,000
Individual Team Sponsorship: Sponsor selects a team to receive funds to offset the cost of building materials, travel, and lodging. Sponsor logo will be displayed next to team listing on the MATE website.	●	●	●	●	●	●
Branding: Sponsor logo placed on marketing materials, on MATE website for 1 year, and pool-side during the competition.		●	●	●	●	●
Tax Benefits: Money, hardware, and/or software contributions to the MATE nonprofit.		●	●	●	●	●
Newsletter & Social Media: Opportunity to announce internships, scholarships, job openings, and other opportunities to faculty and students in MATE's newsletter and on social media.			●	●	●	●
Public Relations: Promote MATE/Sponsor relationship with joint PR opportunities throughout the competition season.			●	●	●	●
Ability to Design Custom Competition Elements: Sponsors collaborate with MATE to create competition scenarios, tasks, props, and more.				●	●	●
Custom Communications: Opportunities to address student competitors with individualized sponsor messages.				●	●	●
Featured Branding: Logo displayed on dedicated section of MATE website, in MATE monthly newsletter, and during Livestream broadcast.				●	●	●
Priority Access: Access to competitors resumes 2 weeks prior to event.					●	●
Inclusive Branding: Logo displayed during award ceremony.					●	●
Exclusive Benefits: Partner with MATE to create benefits to meet your organization's needs.						●
Collect Feedback: Ability to survey competitors and teacher/mentors.						●

For more information, or to discuss custom Strategic Partnership opportunities, contact
Jill Zande, MATE II President and Competition Coordinator, at jzande@marinetech.org

MARINETECH.org

MATE Inspiration for Innovation's tax exempt status under Internal Revenue Code Section 501(c)(3) is pending. Tax I.D.: 81-4389131. Contact jzande@marinetech.org for more information.

MATE II

INSPIRATION FOR INNOVATION
Exhibit Hall

MATE INTERNATIONAL ROV COMPETITION
JUNE 23 THROUGH 25, 2017 | LONG BEACH, CALIFORNIA

SPONSORS: HIGHLIGHT TECHNOLOGIES AND CAREERS

Sponsors embrace this chance to share their latest innovations and career opportunities with more than 800 students and faculty from around the world that attend the MATE International ROV Competition.

BOOTH SPECS
Space Size: 10' x10'
Tables Provided: One 6ft. table
Chairs Provided: Two chairs*
WiFi & electricity provided*

*Extra chairs upon request
*Bring power strip

SHIPPING INFO
Scott Fraser
Long Beach City College
Electrical Technology Department
4901 E. Carson St., Long Beach, CA 90808
MATEIIexhibithall@marinetech.org

***Important:** Please include **"MATE ROV Competition"** on shipping label.

SET-UP/DOWN
Exhibit Set-up Date: June 23
Time: 11:00 AM
Exhibit Breakdown Date: June 25
Time: 11:00 AM

For more information, or to discuss custom Strategic Partnership opportunities, contact Jill Zande, MATE II President and Competition Coordinator, at jzande@marinetech.org

MATE Inspiration for Innovation's tax exempt status under Internal Revenue Code Section 501(c)(3) is pending. Tax I.D.: 81-4389131.

Future directions

- What about the future of MATE II's relationship with Monterey Peninsula College?
- What about "program income" and the fate of the SeaMATE store?

To be continued…

Questions?

Jill Zande
MATE Center/MATE II
jzande@marinetech.org
(831) 646-3082
www.marinetech.org

Competition FB page: @materovcompetition
Twitter: @matecenter
#MATE2017

Micro Underwater Robot
Educational AUV kit

Sergey Mun
CEO, The Center for Robotics Development
Vladivostok, Russia

Robotics Competitions reach

WRO (World Robotics Olympiad)

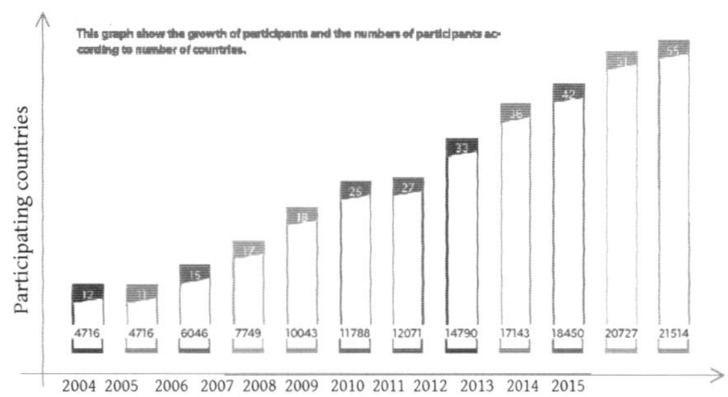

MATE ROV Competition Team Participation 2001-2016

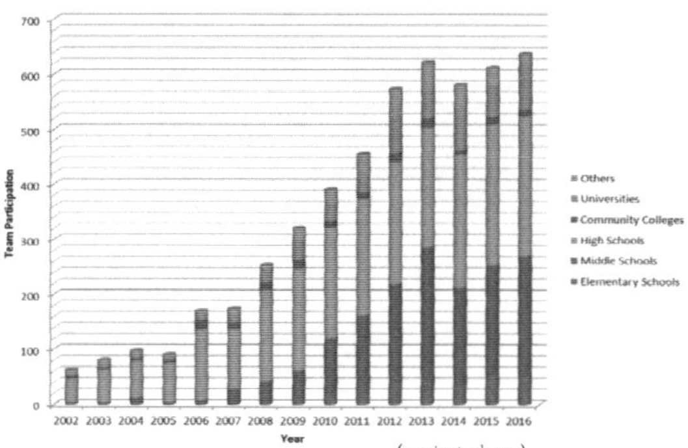

(marinetech.org)

2

High barrier to entry of ROV/AUV Competitions

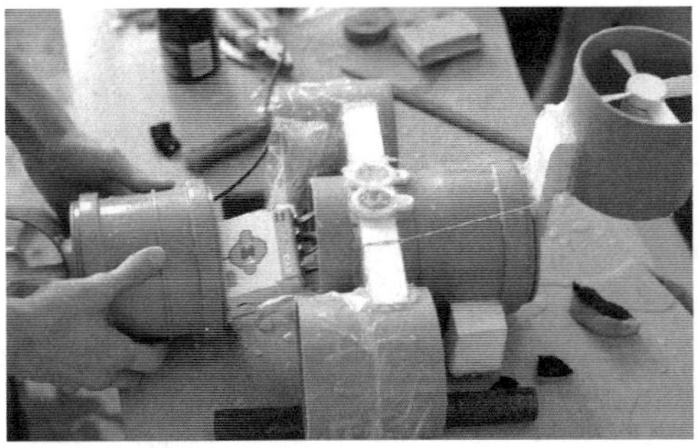

3

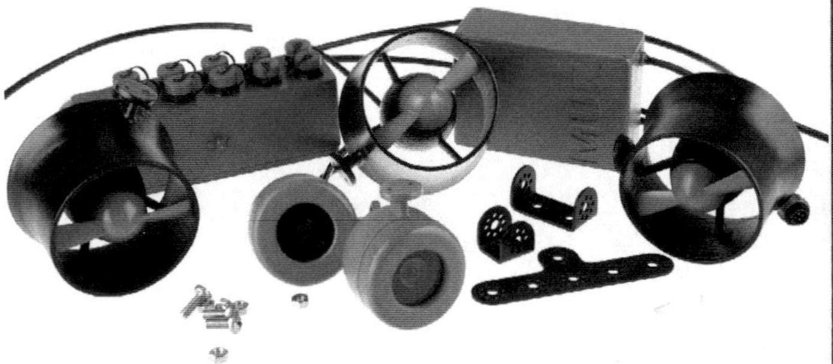

MUR provides

students and makers
with the ability to
DESIGH
And program
UNDERWATER ROBOTS

12+

4

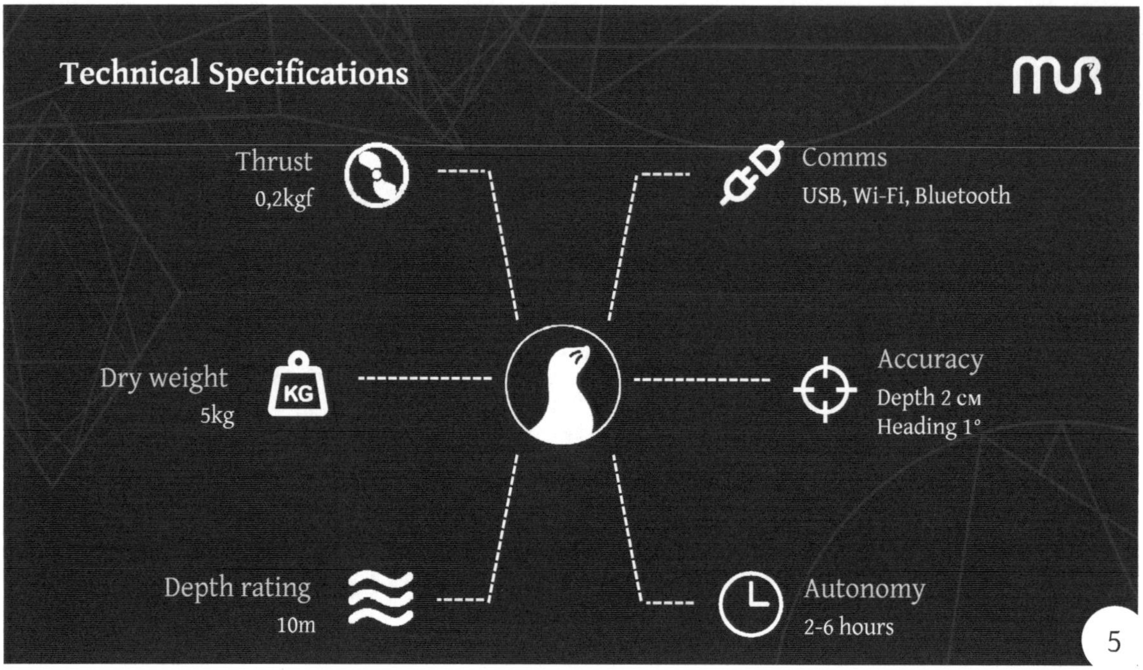
Technical Specifications
MUR
Thrust
0,2kgf
Comms
USB, Wi-Fi, Bluetooth
Dry weight
KG
5kg
Accuracy
Depth 2 см
Heading 1°
Depth rating
10m
Autonomy
2-6 hours
5

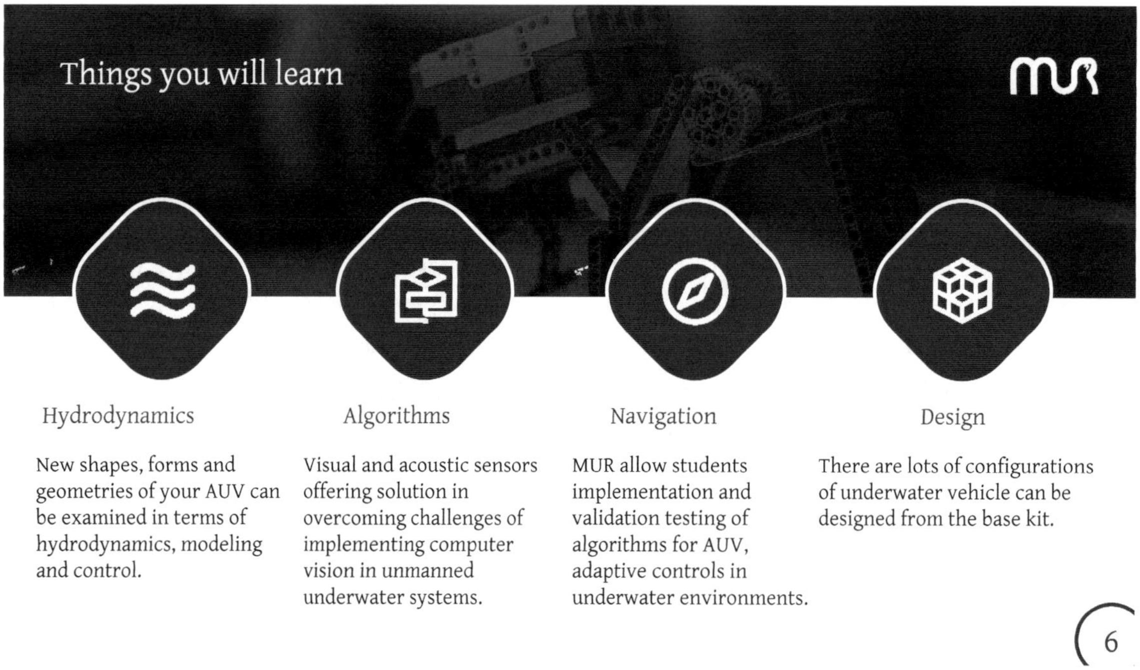
Things you will learn
MUR
Hydrodynamics
New shapes, forms and geometries of your AUV can be examined in terms of hydrodynamics, modeling and control.
Algorithms
Visual and acoustic sensors offering solution in overcoming challenges of implementing computer vision in unmanned underwater systems.
Navigation
MUR allow students implementation and validation testing of algorithms for AUV, adaptive controls in underwater environments.
Design
There are lots of configurations of underwater vehicle can be designed from the base kit.
6

IDE

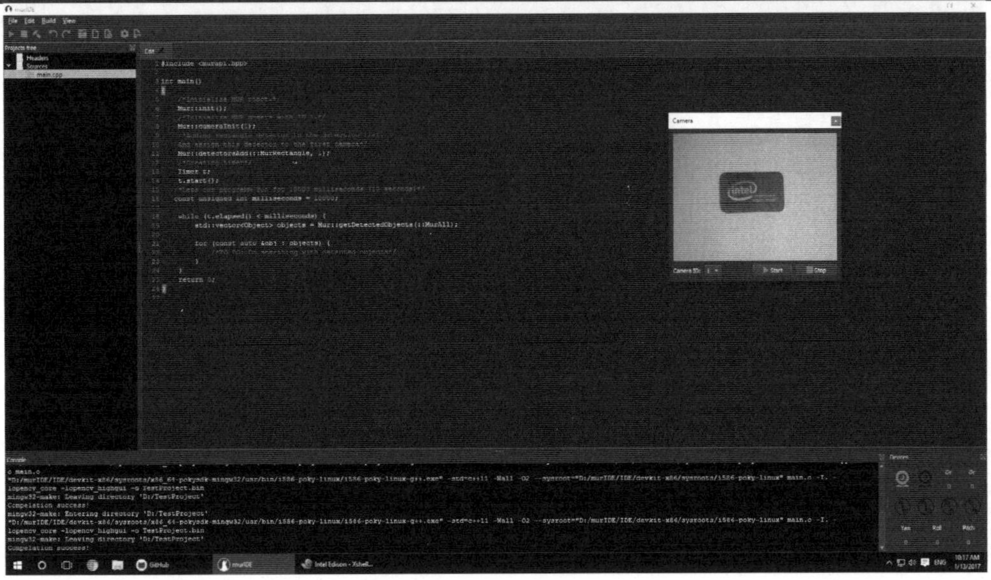

7

Video Presentation in Original e-media

Video
https://vimeo.com/166451884

Russian robotics Olympiad (part of WRO)
AUV Category

- First time in Russia
- 6 teams from Moscow, Kazan, Ufa and Vladivostok
- Teams with MUR won 1, 2 and 3 places

9

Robosub 2017

10

QUESTIONS?

Underwater Technology for Undergraduates: Rapid Design/Build Projects

CDR Andy Gish, PhD, PE
US Naval Academy

Underwater Intervention '17
New Orleans, LA

1

Presentation Outline

- Background and motivation
- Overview of Ocean Engineering at USNA
  - Lab and field facilities
- Rapid design/build project
- Specific project examples from a recent class :
  - ROV Payload Delivery Module
  - Rotating ROV Manipulator
  - Magnetic Motor Coupling
- Discussion

2

Background and Motivation

- Underwater Technology is part of many undergrad engineering programs
  - Traditional engineering courses do not address unique underwater challenges
  - Shortage of good textbooks on the subject
  - Shortage of interested and qualified faculty
- Project-based learning (PBL) is effective and popular with students
  - Downside: PBL can be expensive and time-intensive

Source: //store.openrov.com/

- **<u>CHALLENGE</u>: How to most effectively use PBL to teach underwater technology principles (design, construction, and operation)?**
- **<u>SOLUTION:</u> Rapid design/build projects**

3

United States Naval Academy Ocean Engineering

- 4000 undergraduate students, 0 graduate students
- 25 different majors (10 engineering)
- Faculty: 50% tenure-track civilian, 50% military officers
- Very rigorous academic/military/physical regimen
- 60-70 Ocean Engineering majors per class
  - 14 required core engineering courses
  - 4 engineering electives (senior year)
    - Underwater track: Underwater Work Systems and Life Support Systems
  - Senior Capstone Design Project (2 semesters)
  - Opportunities for independent/honors research projects (1 or 2 semesters)

4

United States Naval Academy
Ocean Engineering

- 4000 undergraduate students, 0 graduate students
- 25 different majors (10 engineering)
- Faculty: 50% tenure-track civilian, 50% military officers
- Very rigorous academic/military/physical regimen
- 60-70 Ocean Engineering majors per class
  - 14 required core engineering courses
  - 4 engineering electives (senior year)
    - Underwater track: Underwater Work Systems and Life Support Systems
  - Senior Capstone Design Project (2 semesters)
  - Opportunities for independent research projects (1 or 2 semesters)

5

United States Naval Academy
Lab Facilities

- **380 foot towing tank:**

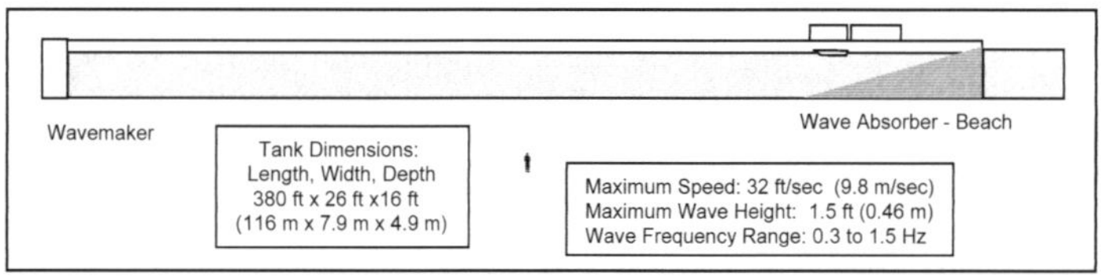

- **120 foot towing tank:**

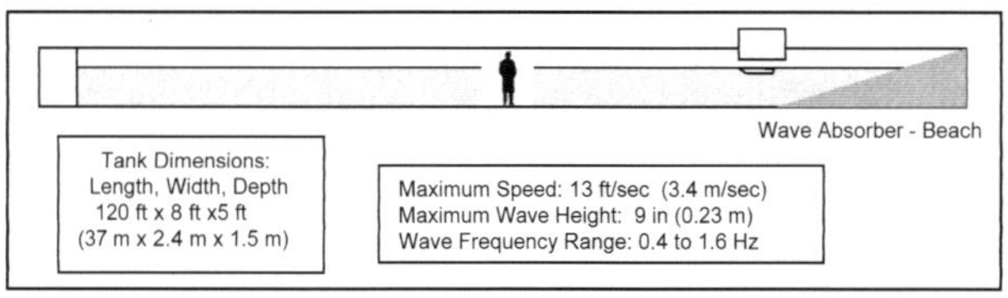

- **Circulating water channel:** max. speed 25 ft/sec
  - test section 16"x16"x5'

- **Coastal engineering basin:** max. depth 1 foot
  - used to study wave effects on beaches and structures
- **Ballast tank:** depth=4 feet
  - used for static tests and ballasting vehicles

6

United States Naval Academy Field Facilities

- Pier/seawall access to Severn River and College Creek
- Small boat/Yard Patrol (YP) craft access to river and Chesapeake Bay

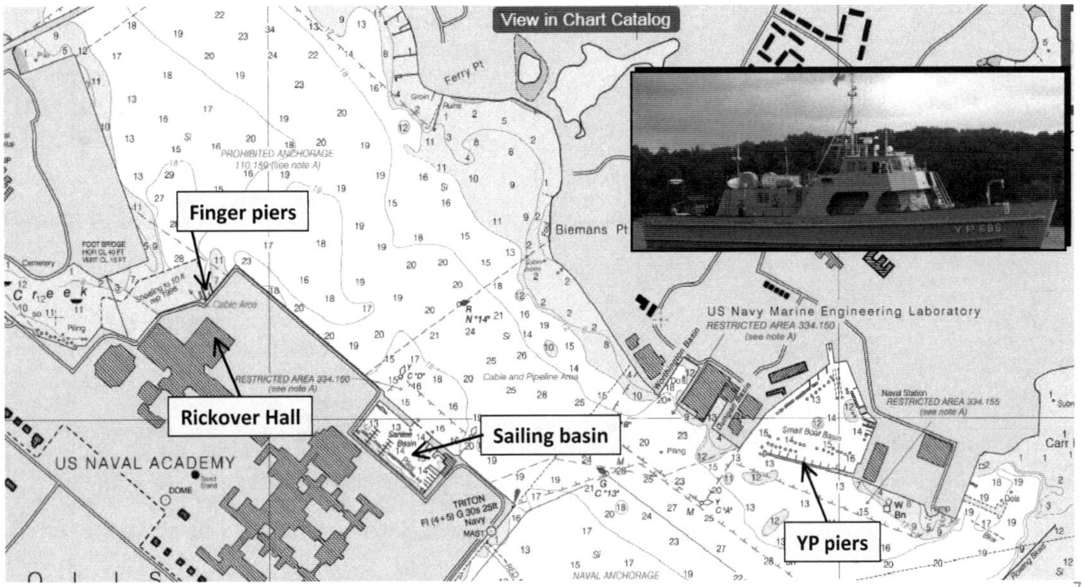

7

Rapid Design/Build Project

- **Objective:** give students practical, hands-on experience in the design, construction, and testing of an underwater system
- **Timeframe**: 2 weeks (at end of semester)
- **Venue:** senior-level Ocean Engineering elective course
- **Assignment:** teams of 4 students
  - **Background/Problem Definition**
  - **Brainstorming/Development of Solutions**
  - **SolidWorks model/sketch of chosen design**
  - **Supporting calculations (hand or spreadsheet)**
  - **Physical Prototype of your design**
  - **Discussion/Results**
  - **Next Steps/Future Work**
- **Constraint:** must utilize materials and components available in the USNA labs or project shop
- **Deliverables:** initial design review, final written report, team presentation
- Project done concurrently with senior Capstone Design project

8

Payload Delivery Module

- **Objective**: Fabricate an ROV-based system to carry a payload object and place it at a designated location on the bottom.
- **Design criteria**:
- Payload W=6 lbs, V=12 in^3
- Near neutrally buoyant
- Add minimal drag while transiting
- Remotely released from the surface
- ROV neutrally buoyant and unencumbered after release
- Reusable
- ROV fully maneuverable with payload

- **Final Design**: Vertical foam pieces sandwiched around payload
- **Operation:**
- Place weight in foam apparatus
- Clamp apparatus with ROV claws
- Drive ROV to payload position
- Release claws
- Deliver payload
- Foam floats to top

- **Analysis of Alternatives**: 3 designs considered

	Weight	'Horizontal' foam system	'Vertical' foam system	Air bag system
Ease of construction	0.50	3	3	1
Hydrodynamic drag	0.40	1	2	3
Reusability	0.10	3	3	1
Score		2.2	2.6	1.8

- **Future Work:**
- Create a way to recollect foam more easily
- Lighter foam to decrease the required volume
- Flexibility for other payload shapes
- **Lessons Learned:**
- Brainstorm excessively prior to beginning project
- Most complex solution not always the best
- Measure twice, cut once
- Record every trial
- Design for repeatability - consider material selection

9

ROV Rotating Manipulator

- **Objective**: Design and fabricate a mechanism to allow rotation of the existing VideoRay Pro4 manipulator
- **Design criteria**:
- Device must be remotely operated from the surface
- Rotate the existing manipulator arm at least 90^0
- Rotate the manipulator both directions

- **Final Design**:
- Waterproof DC Servo
- Gear attachment on manipulator shaft
- Driven by MBED microprocessor

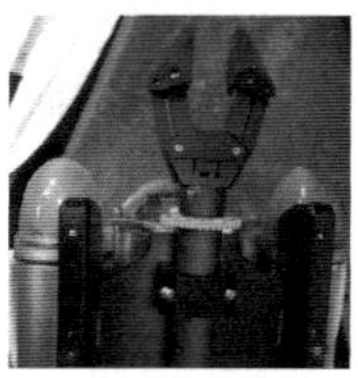

- **Analysis of Alternatives**:
- Belt drive or Gear drive?
- Rotate entire manipulator shaft or just end?
- Wireless or wired operation?

- **Future Work:**
- Stepper motor instead of a servo motor
- Metal gears instead of plastic
- Integrate with existing tether and controls
- **Difficulties/Lessons Learned:**
- Mounting the gear to the shaft
- Waterproofing the wire connection
- Keeping the shaft and manipulator arm positioned forward and aft
- Turning the shaft full 90^0

10

ROV Rotating Manipulator

Video Presentation in
Original e-media

(Video)

11

Magnetic Motor Coupling

- **Objective**: Design and fabricate a working magnetic coupling for a standard SeaPerch DC motor.
- **Design criteria**:
- Operable to at least 12 ft depth
- Use available rare-earth magnets (0.25”x0.25”x0.1”)
- 3D print or fabricate in USNA shop
- **Analysis of Alternatives**:
- Concentric coupling rings vs. face-face coupling
- Optimal number of magnets?
- Alternate poles (N/S) or not?
- Fabricate housing or use available canister?
- How to attach prop

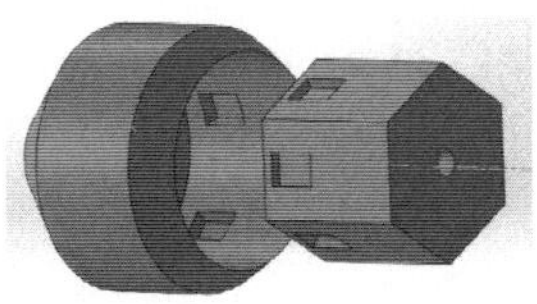

- **Final Design**:
- Concentric rings
- 6 magnets per ring, alternating poles
- Standard pill bottle used for housing
- Motor press-fit inside housing

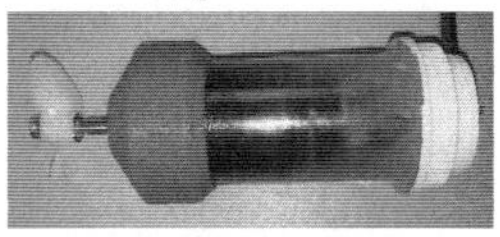

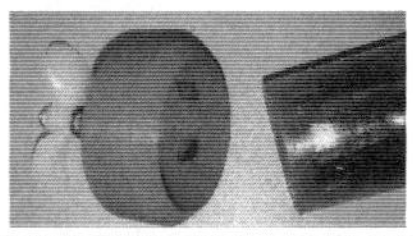

- **Future Work:**
- Use bearings to reduce friction
- Redesign inner ring to reduce air gaps
- Reduce mass of outer ring
- Attach prop blades directly to outer ring
- **Lessons Learned:**
- It works! (in air and water, all orientations)
- Magnets are very strong, difficult to attach
- Water lubrication reduces friction

12

Magnetic Motor Coupling

Video Presentation in Original e-media

(Video)

13

Student Feedback

- Overall response : overwhelmingly positive!
- Specific student comments (raw and unedited):
  - I think the most effective learning experience was the project at the end. You could really see the actual application of the course and how the stuff we learned applied to real life scenarios.
  - The end of class project was a very helpful learning experience because we were forced to go through the entire process from ideas to design to an actual prototype.
  - This was for me the first time I had actually had something I made in Solidworks be 3D printed which was really cool.
  - The rapid design project was the most effective because it allowed us some freedom to work on a tangible project which we could see through until we finished and we could take pride in it. The rapid design project is what I expected when I chose Ocean Eng, so I am glad I finally got to do something hands on.

14

Discussion

- Questions?
- Suggestions to improve or expand these examples?
- Other project ideas?

Contact information:
CDR Andy Gish
U.S. Naval Academy
410-293-6422
gish@usna.edu

15

PATHWAYS TO STEM THROUGH UNDERWATER ROBOTICS

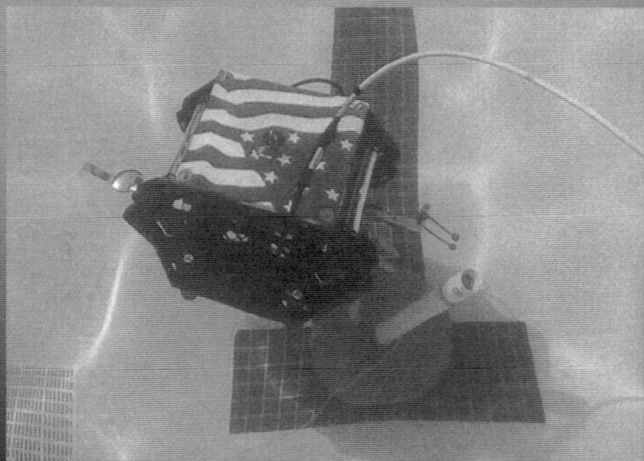

COPIAH-LINCOLN COMMUNITY COLLEGE

- Serves a 7 county region in SW-MS
- 3 Campuses
- 670 Borders
- 3,742 full and part time students
- (White 53%Black 42%
- 59% Pell Grant
- 68 % receive fi[illegible] aid

Pop.154,161
~Ft. Lauderdale

DEMOGRAPHICS

*	Lincoln County	Jefferson County	MS	USA
%White	68.3%	13.4%	60.6%	79.8%
%Black	30.7%	86.0%	37.2%	12.8%
%HS	72.0%	59.7%	72.9%	80.4%
%College	12.4%	10.9%	16.9%	24.4%
% Below Poverty	18.4%	35.2%	20.7%	13.0%
5th Grade Science	566.2	542.2	556.5	N/A

*http://quickfacts.census.gov

HOW TO ENCOURAGE STEM IN A COMMUNITY

SUMMER CAMPS

FIRST LEGO LEAGUE

FIRST TECH CHALLANGE

MS TRENDS IN CTE

- Mississippi's new science standards integrate engineering design, bringing it alongside the scientific method as a problem solving method
- The majority of CTE Engineering students will NOT go on to be engineers, but all students deserve access to this program and way of thinking

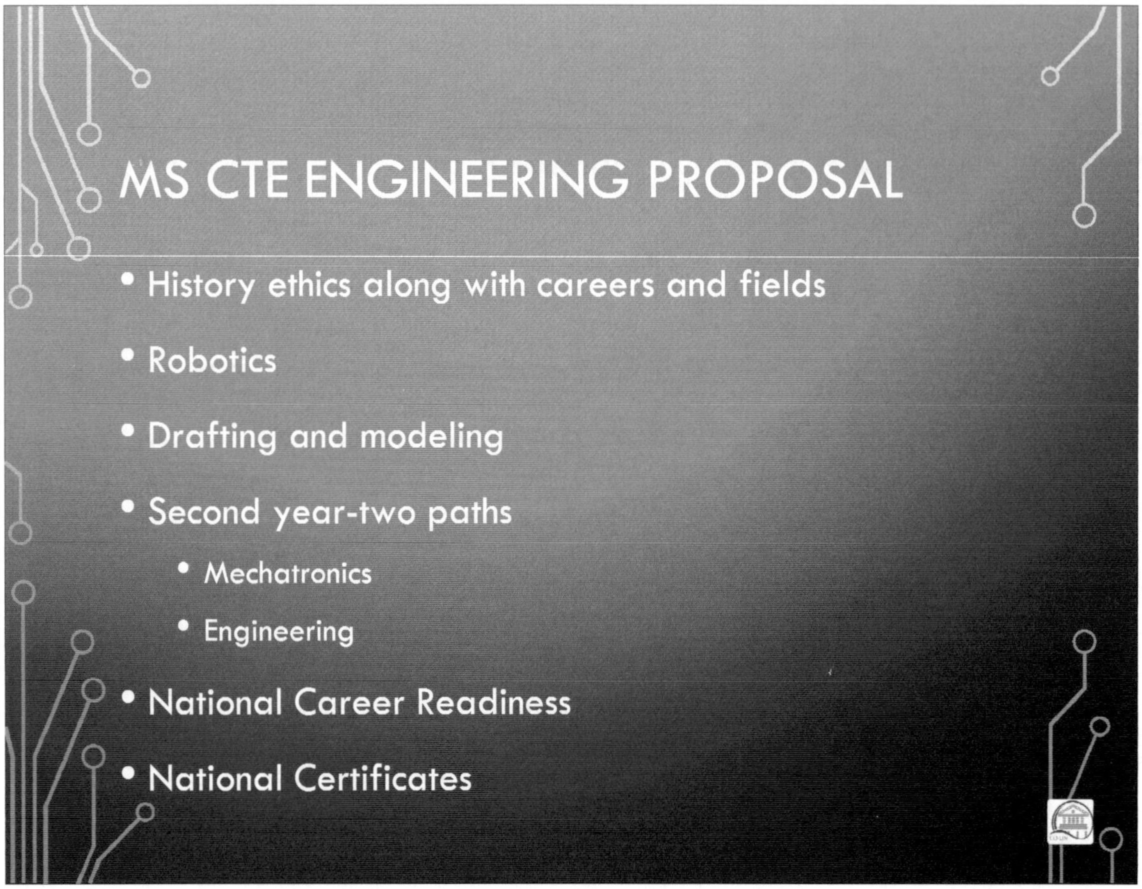
MS CTE ENGINEERING PROPOSAL
• History ethics along with careers and fields
• Robotics
• Drafting and modeling
• Second year-two paths
• Mechatronics
• Engineering
• National Career Readiness
• National Certificates

MATE INTERNATIONAL ROV COMPETITION
16th Annual
MATE International ROV Competition
Port Cities of the Future:
Commerce, Entertainment,
Health, and Safety
MTS
OCEANEERING
LONG BEACH
June 23-25, 2017 » Long Beach City College » Long Beach, California

2017 DESIGN

NATIONAL BUOYANCY LAB 2016

SKILLS

PRECISION MACHING

DRAFTING AND DESIGN

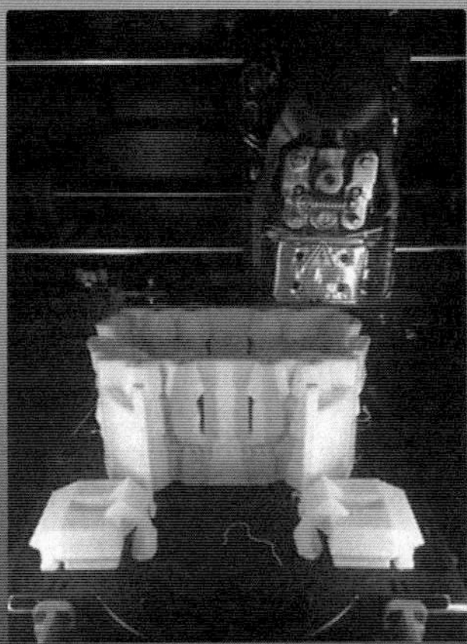

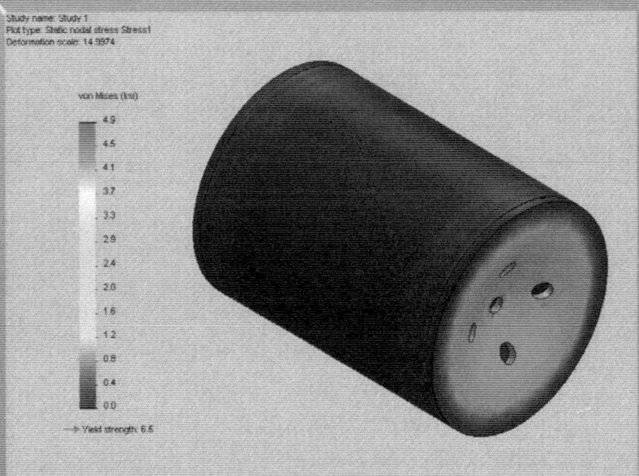

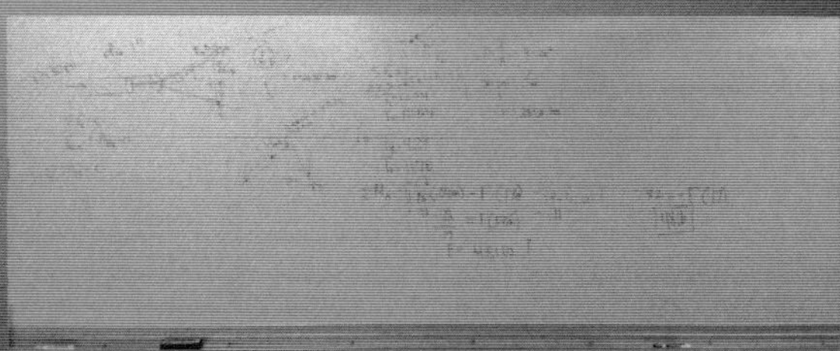

ENGINEERING

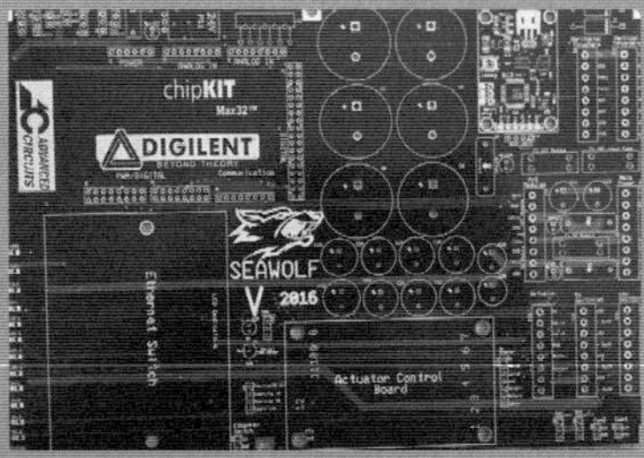

ELECTRONICS

CAREERS

NISSAN

I-55 ENGINEERING

GEORGIA PACIFIC

CAREER PLACEMENT

- Eaton Aerospace
- General Dynamics
- SeaTrepid
- C -Innovations
- Oceaneering

MENTORS

- BO JOHNSON- PRECISION MACHINING
- WES BURKETT- CAD
- CAREY WILLIAMSON-ELECTRONICS
- KEVIN MCKONE- PHYSICS
- CELESTE WILLIAMSON-GRAPHICS DESIGN

THANKS!

- Co-Lin Foundation
- Co-Lin Administration
- MATE
- SeaTrepid
- Outland Tech.
- Georgia Pacific
- Spectrum (API)
- Brookhill on Natchez
- Walt Holified I-55 Engineering

WHAT IS THE OBSESSION WITH ROV HORSE POWER AND AGE, THINK TASK

Underwater Intervention 2017

Chris Roper

Saab Seaeye Ltd

Contents

Forget HP, Age & Class

Think TASK

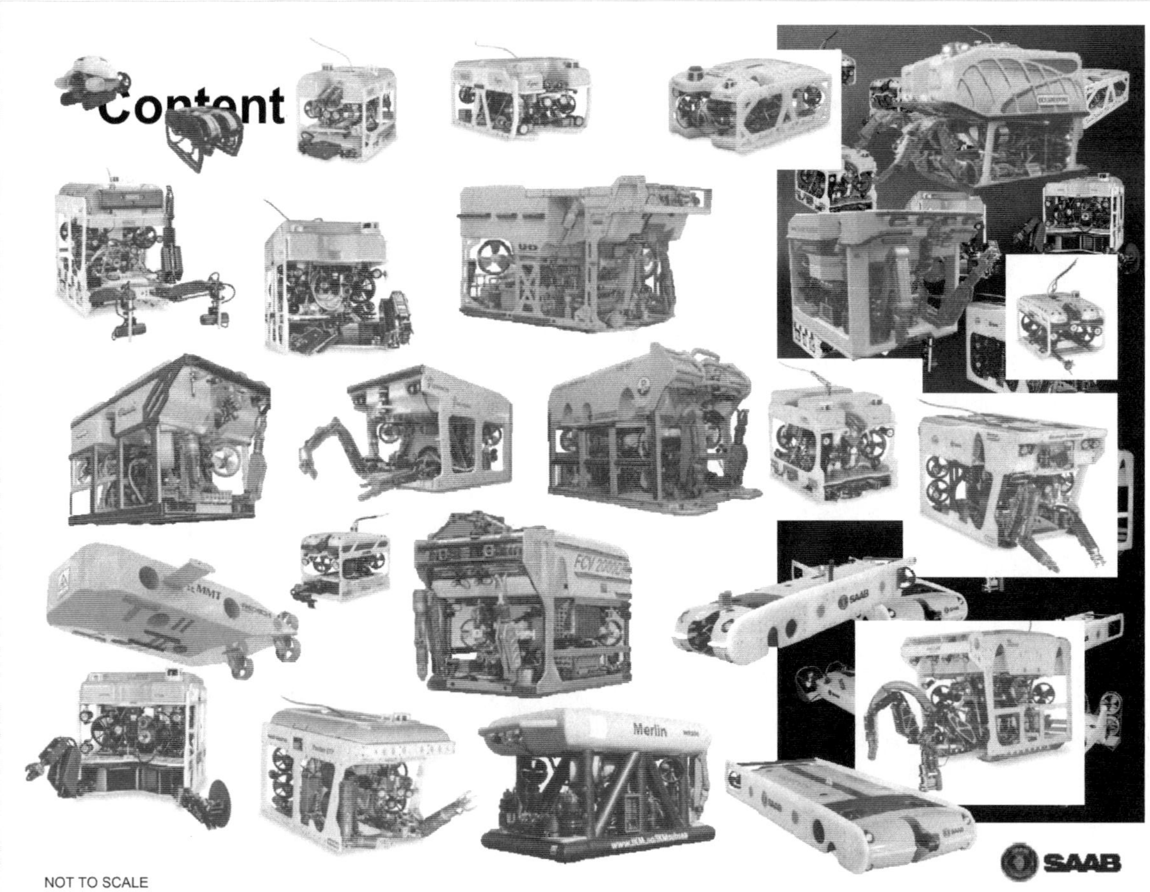
Content
Merlin
NOT TO SCALE
SAAB

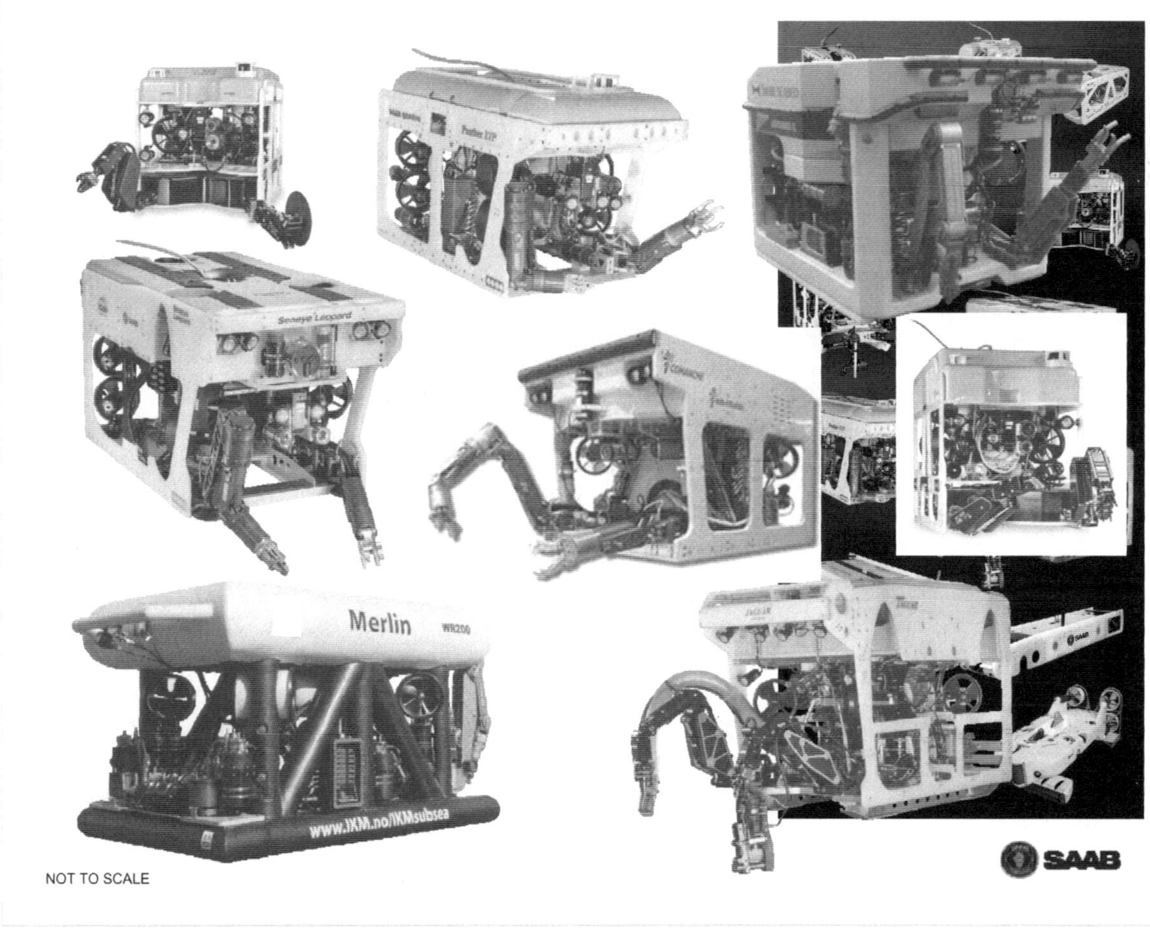
Merlin
WR200
www.IKM.no/IKMsubsea
NOT TO SCALE
SAAB

SAAB

Contents

Progression of Hydraulic ROV capability

	50HP	75HP	100HP	150HP
Weight	1.3 - 1.7 Te	1.5 - 2.1 Te	1.5 - 3.25 Te	[illegible]
Volume	3 - 4 m^3	3.5 - 10 m^3	4.5 - 11 m^3	[illegible]
Payload	70-100 kg	100-200 kg	150-300 kg	[illegible]

SAAB

Contents

The problem of defining ROVs with HP

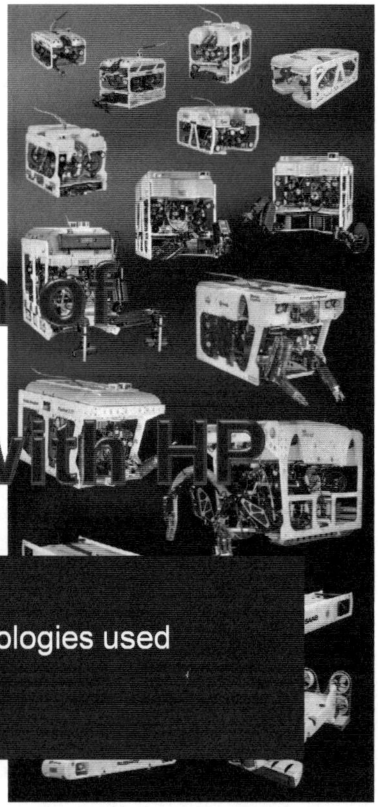

- Can't compare all ROV designs & technologies used
- Power costs

SAAB

7

Contents

Horse Power in Hydraulic Systems

SAAB

8

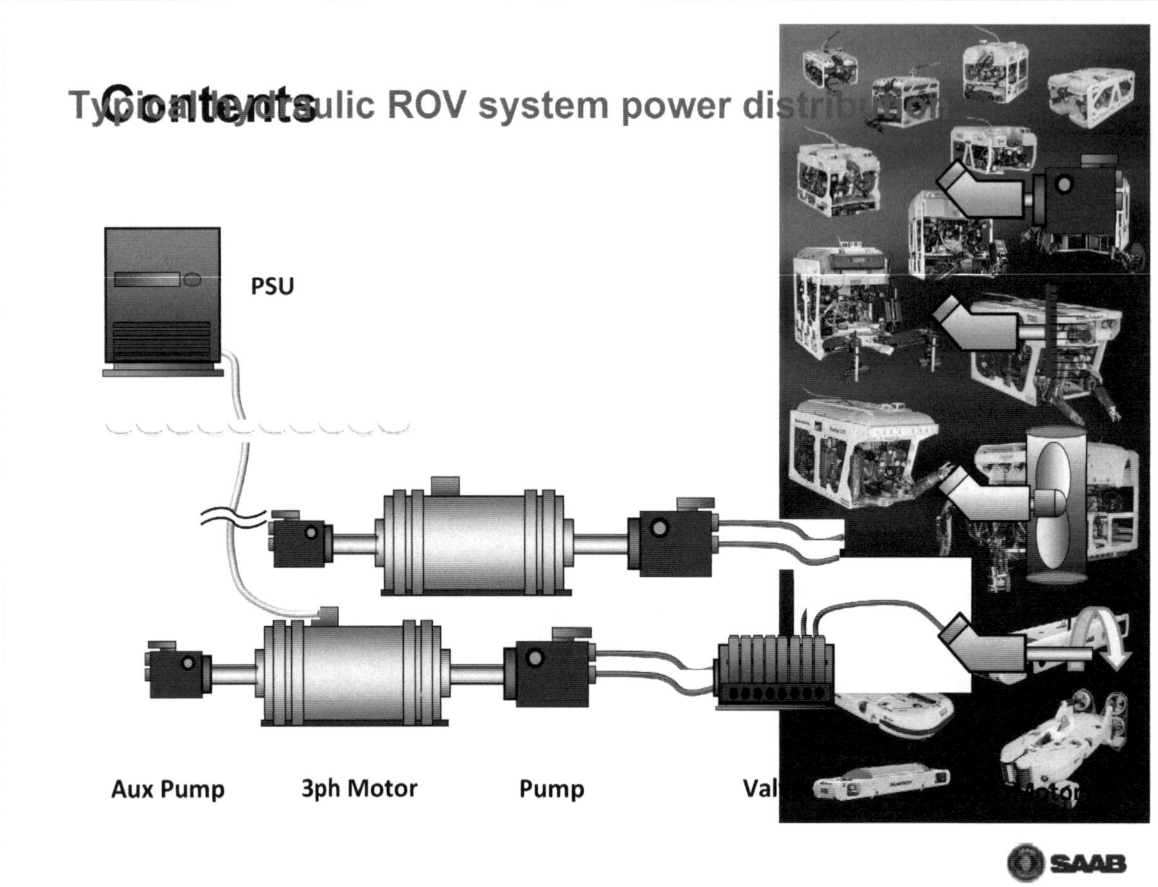
Contents
ulic ROV system power di
PSU
Aux Pump
3ph Motor
Pump
Val
SAAB

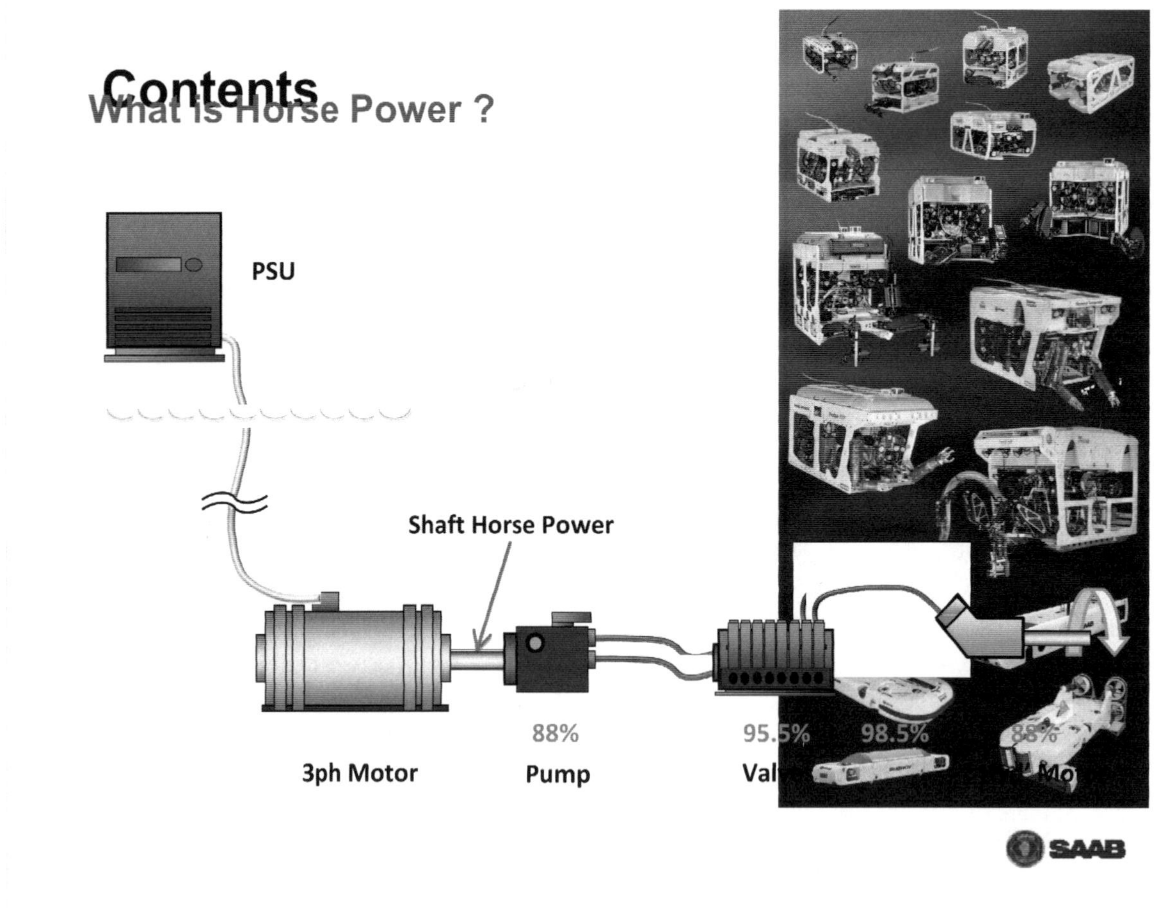
Contents
e Power ?
PSU
Shaft Horse Power
88%
95.5%
98.5%
3ph Motor
Pump
Val
SAAB

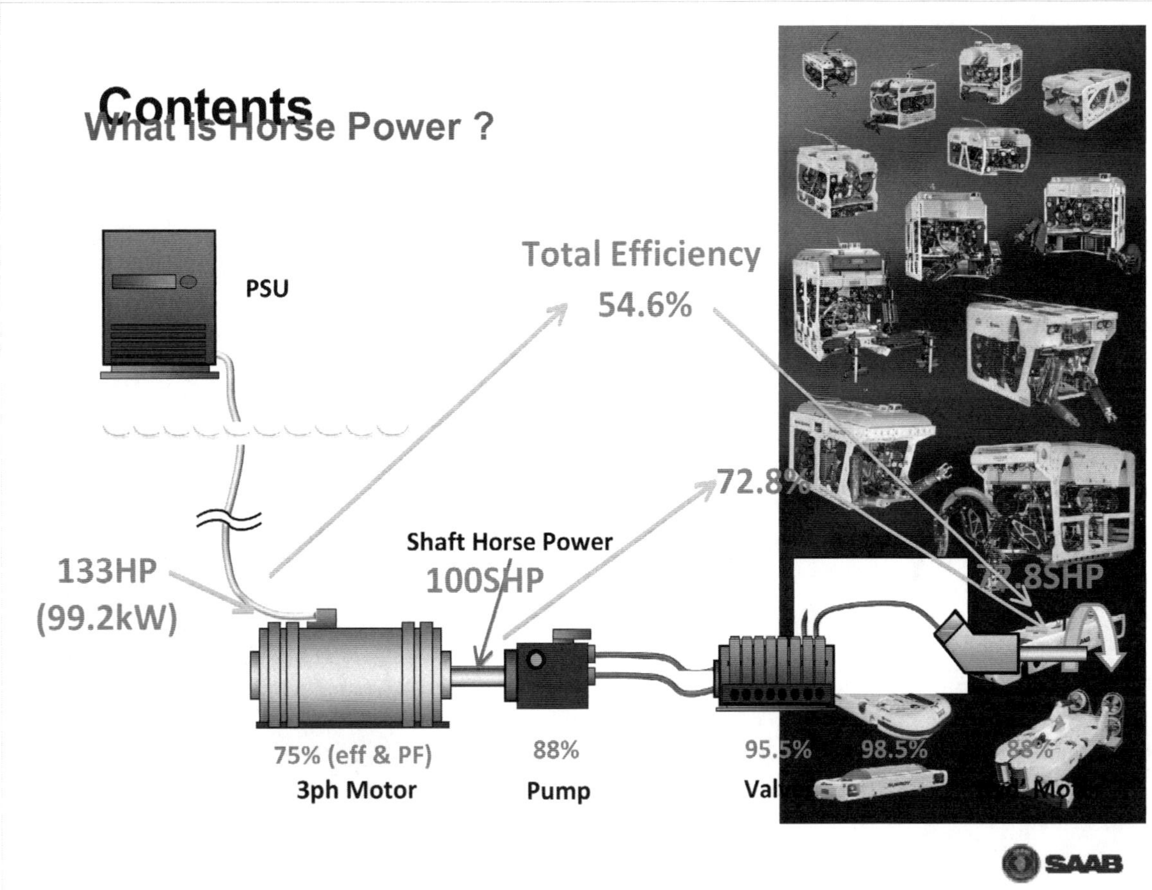
Contents
What is Horse Power ?
PSU
Total Efficiency
54.6%
72.8%
Shaft Horse Power
100SHP
133HP
(99.2kW)
75% (eff & PF)
3ph Motor
88%
Pump
95.5%
98.5%
88%
SAAB

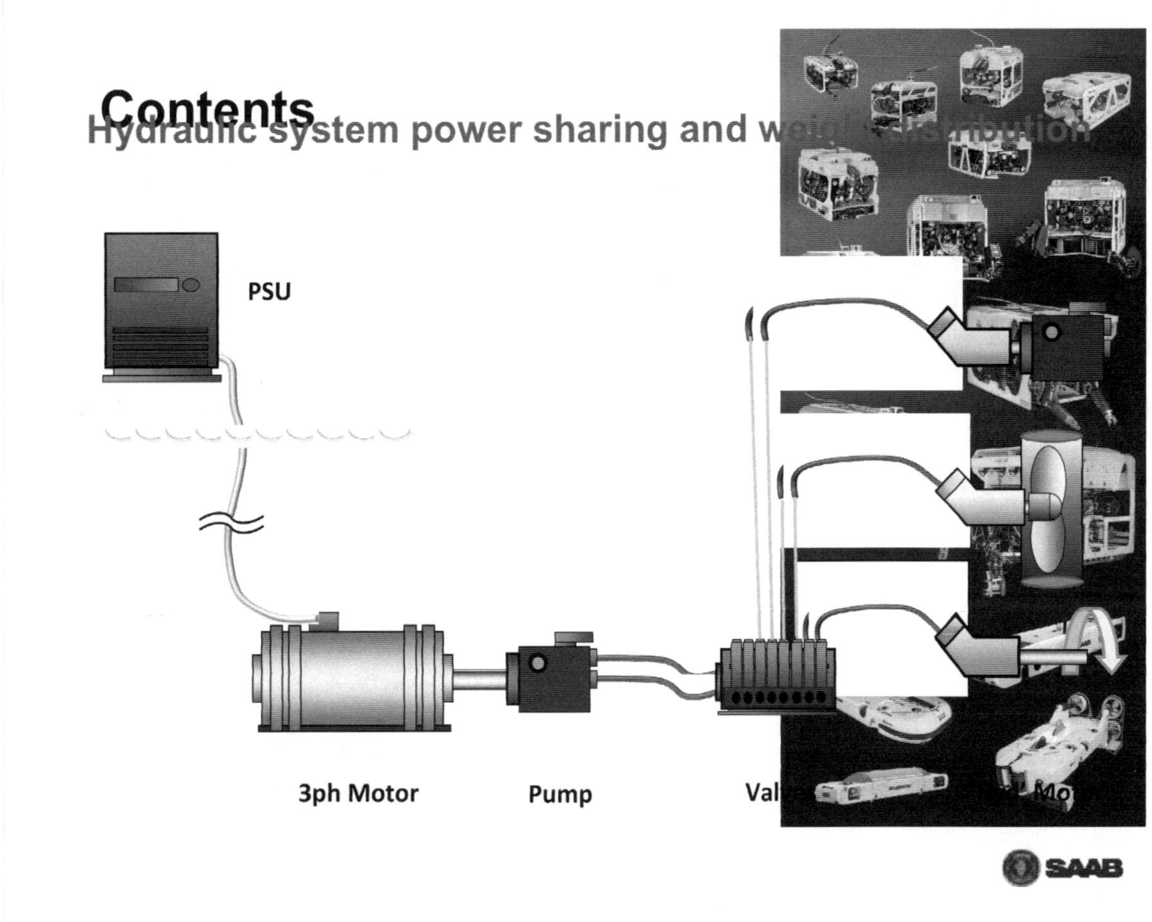
Contents
Hydraulic system power sharing and
PSU
3ph Motor
Pump
SAAB

Contents

Horse Power in Electric Systems

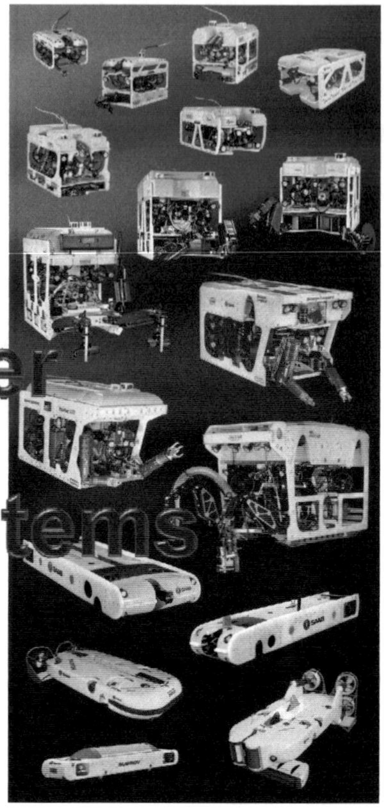

Contents

Typical electric ROV power distribution

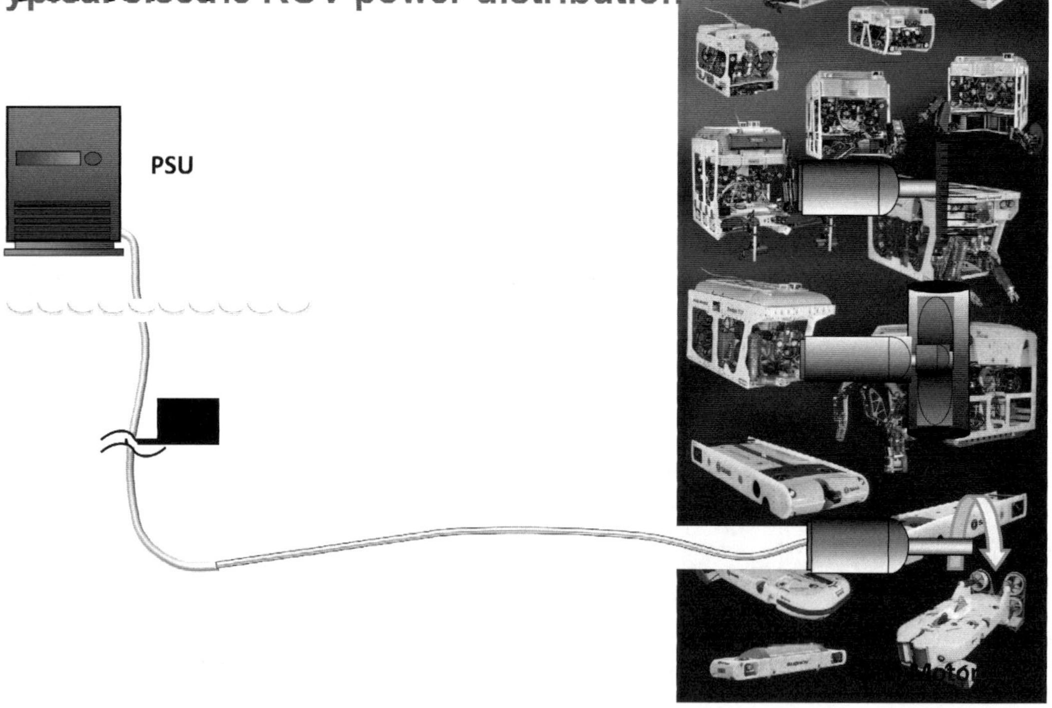

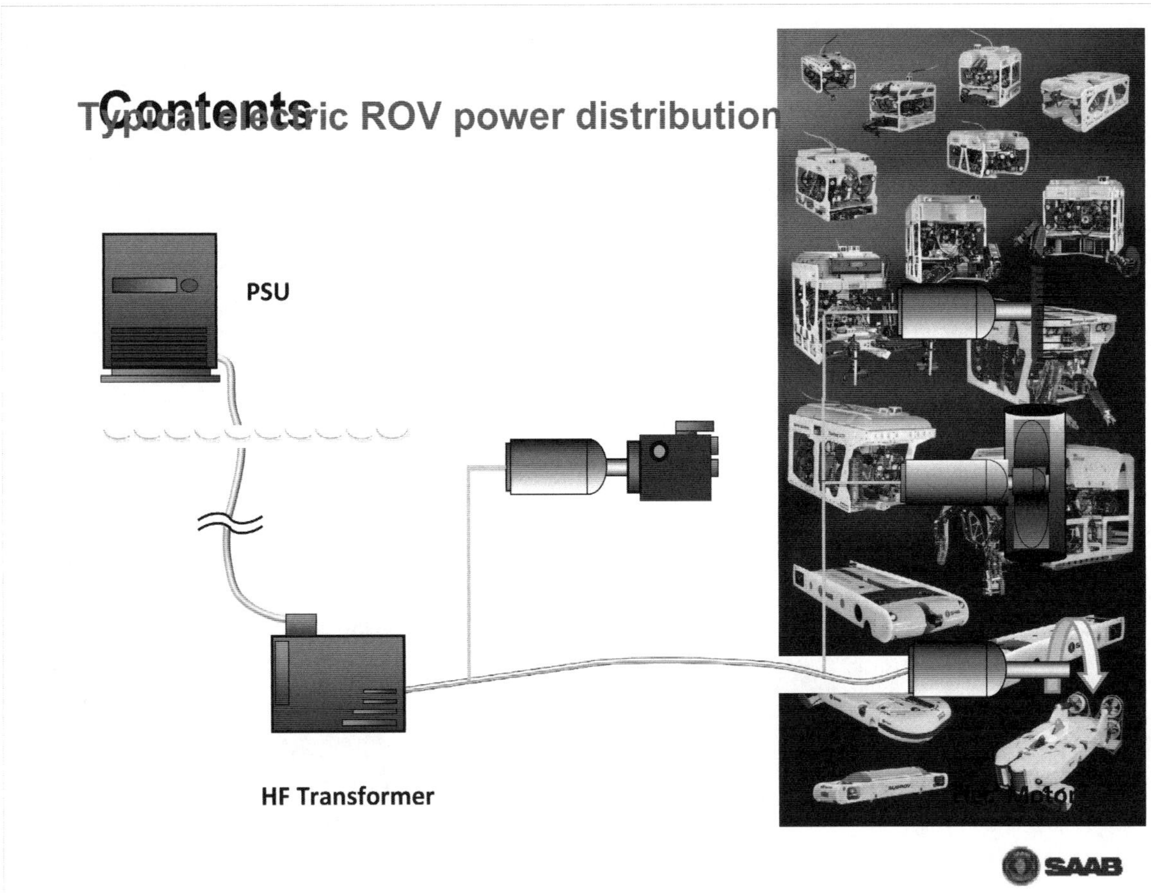
Contents
Typical electric ROV power distribution
PSU
HF Transformer

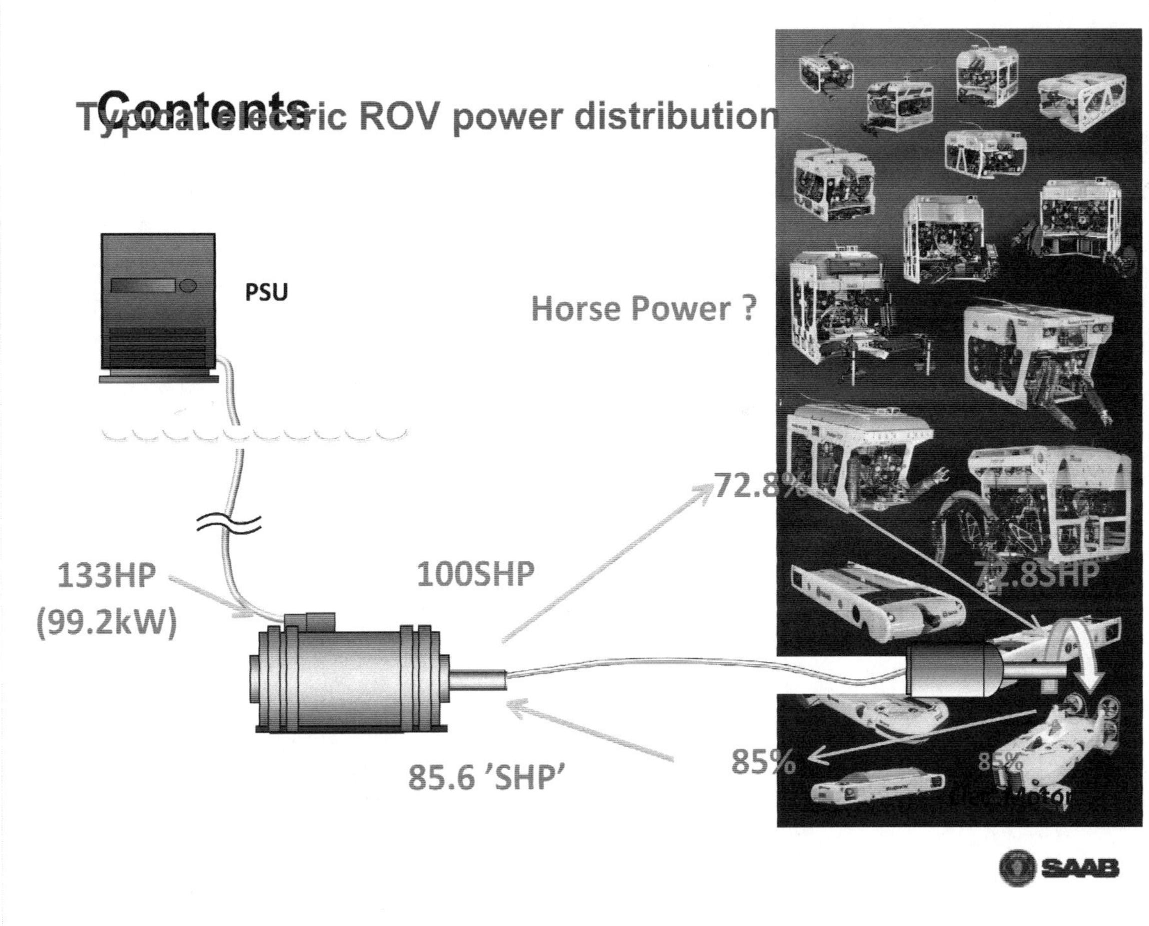
Contents
Typical electric ROV power distribution
PSU
Horse Power ?
133HP
(99.2kW)
100SHP
72.8%
72.8SHP
85%
85%
85.6 'SHP'

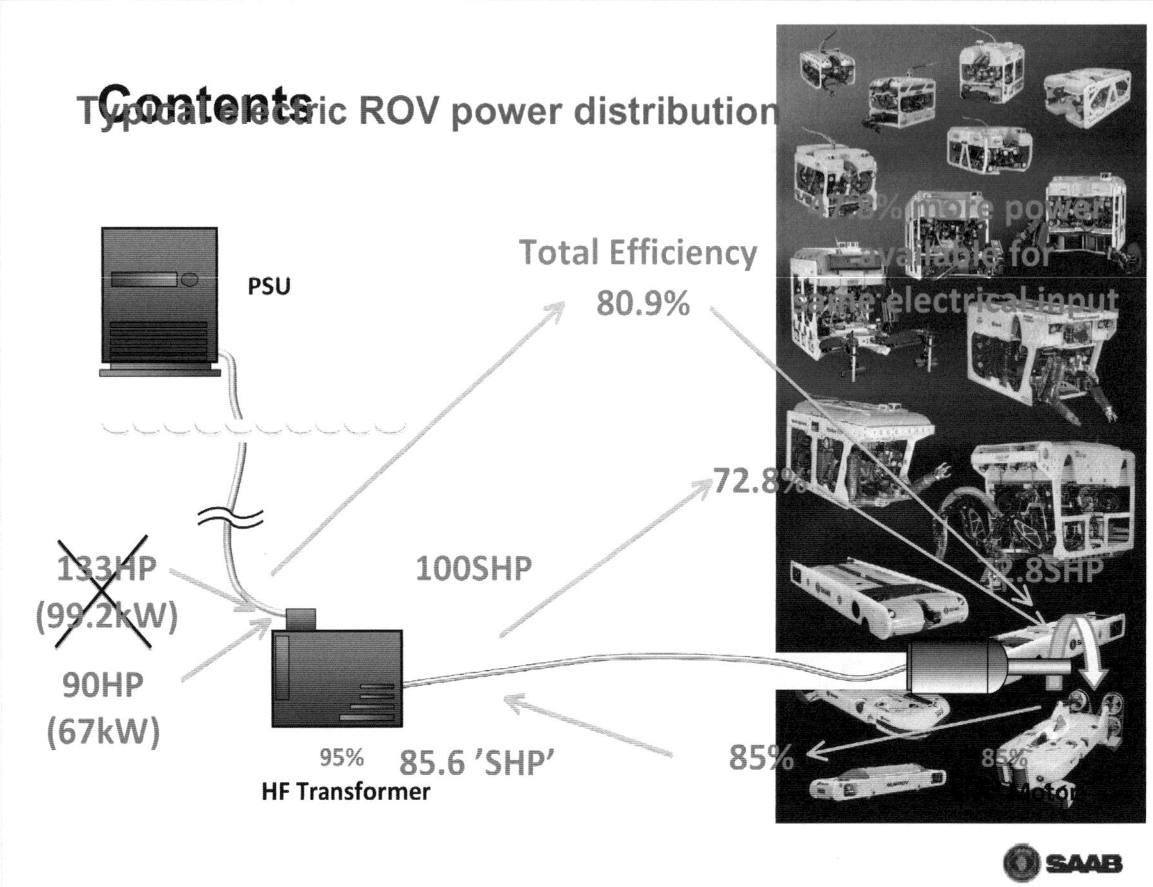
Typical electric ROV power distribution
PSU
Total Efficiency
80.9%
72.8%
133HP
(99.2kW)
100SHP
90HP
(67kW)
95%
85.6 'SHP'
HF Transformer
85%
SAAB

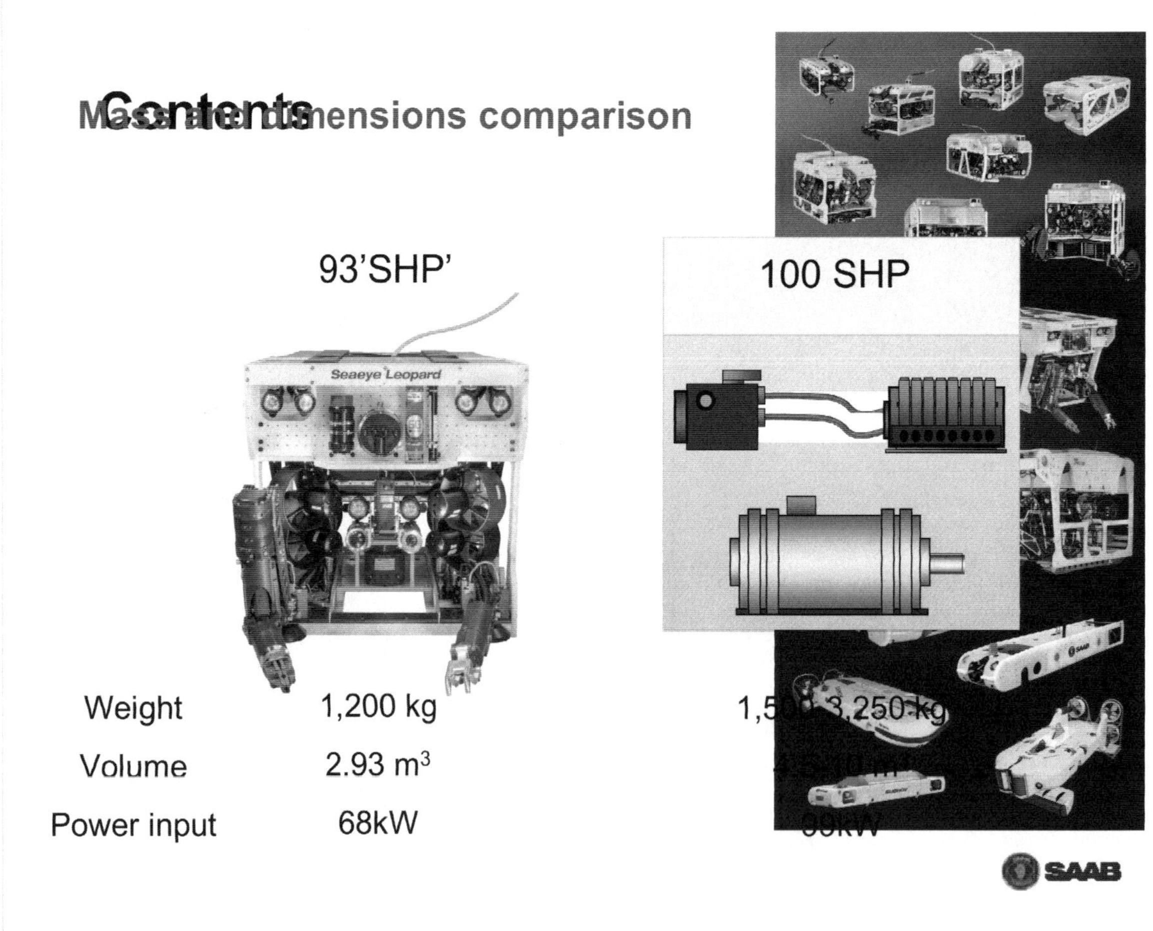
Mass and dimensions comparison
93'SHP'
100 SHP
Seaeye Leopard
Weight
1,200 kg
Volume
2.93 m³
Power input
68kW
SAAB

Contents

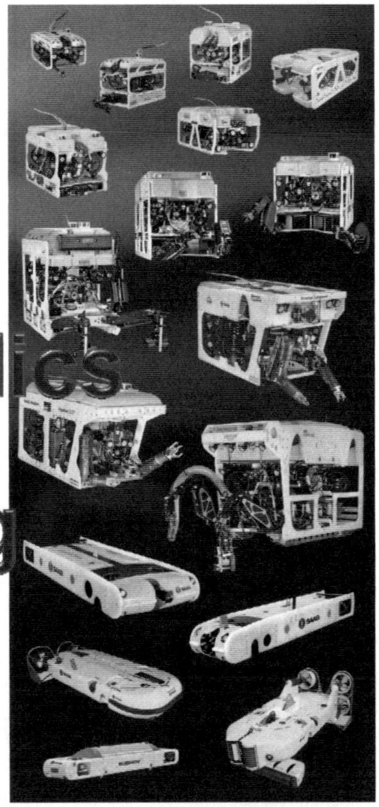

Need hydraulics for tooling

SAAB

Contents

SAAB

Credit - Fugro

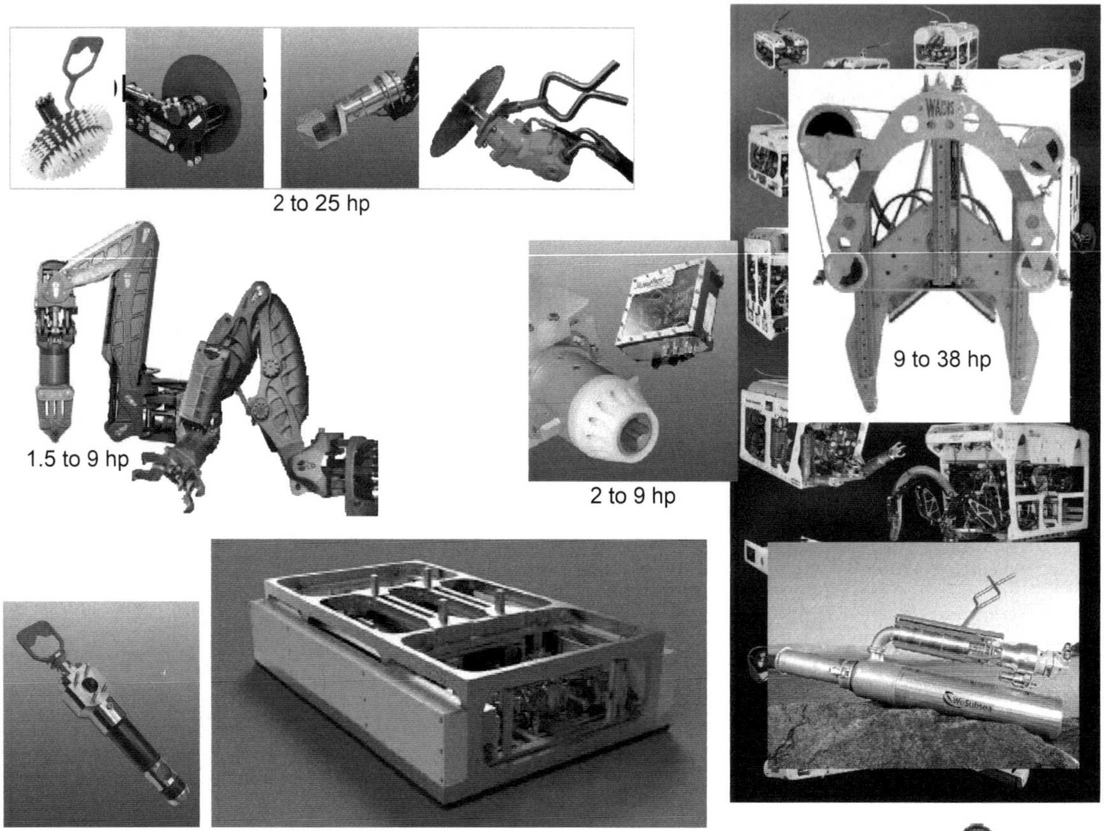

SAAB

Contents

Benefits of Hydraulic and Electric Vehicles

HYDRAULIC VEHICLES	ELECTRIC VEHICLES
Share power between propulsion and tooling & generally greater tooling power	Efficient – can handle strong currents better & generates less heat
Power distribution assists weight and positioning on vehicle	Compact & light weight – access to structures & smaller deck footprint
Larger vehicles able to carry more tooling in one dive	Acoustically quieter – better for survey applications
Availability of tooling in market, designed with the hydraulic ROV in mind	Environmentally friendly (less oil and power)
Well established operating standards and interfaces	More cost effective to purchase and operate

SAAB

Contents

Important factors

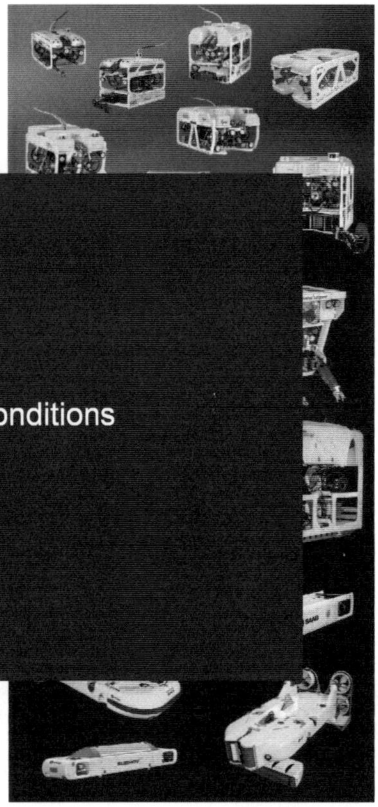

- Equipment to be carried
- Tooling power needed
- Required interfaces
- Expected Operational / environmental conditions
- Deployment platform
- Available skills
- Project duration
- What are the tasks ?

SAAB

Contents

Power Costs

SAAB

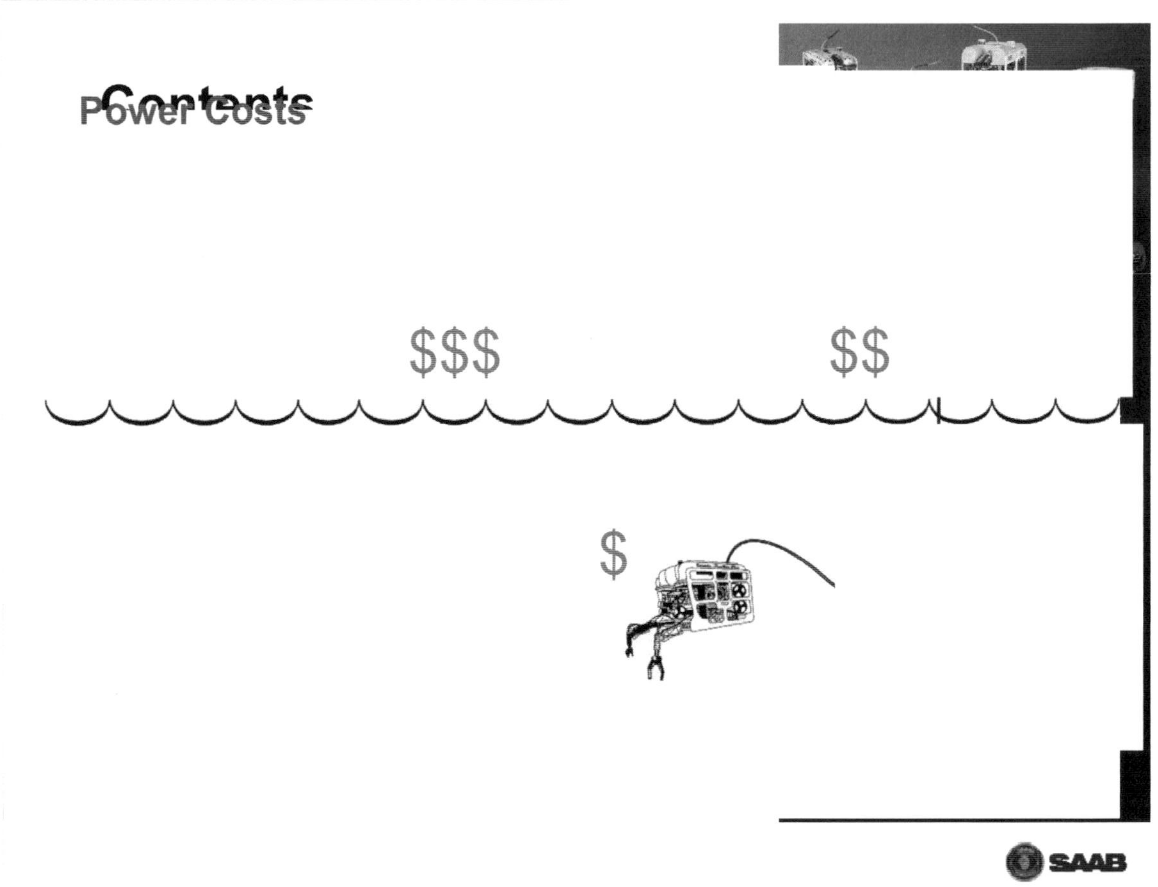
Contents
Power Costs
$$$
$$
$

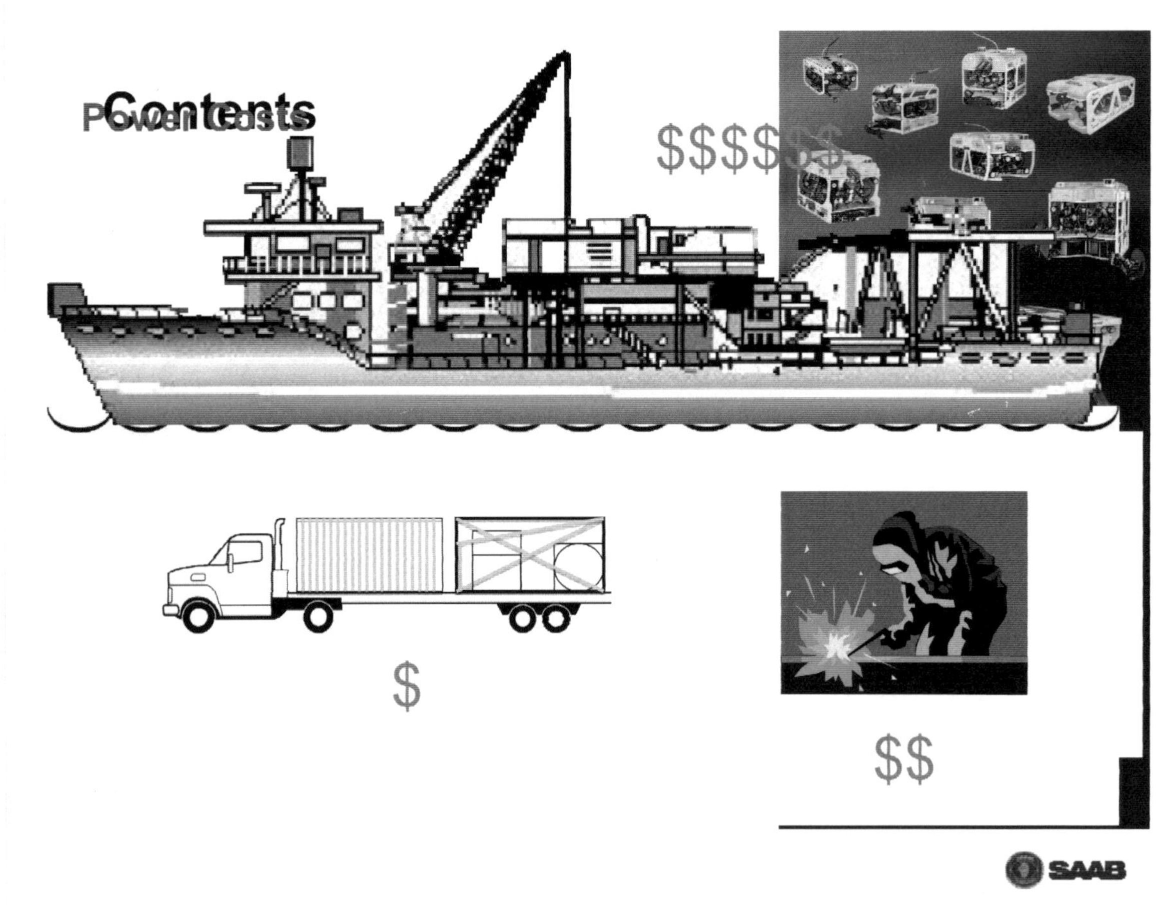
Contents
Power Costs
$$$$
$
$$

Contents

What about Class ?

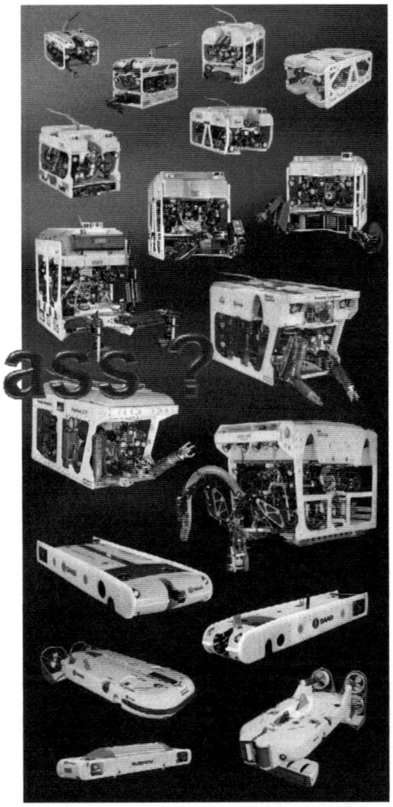

Contents

Typical ROV Classes

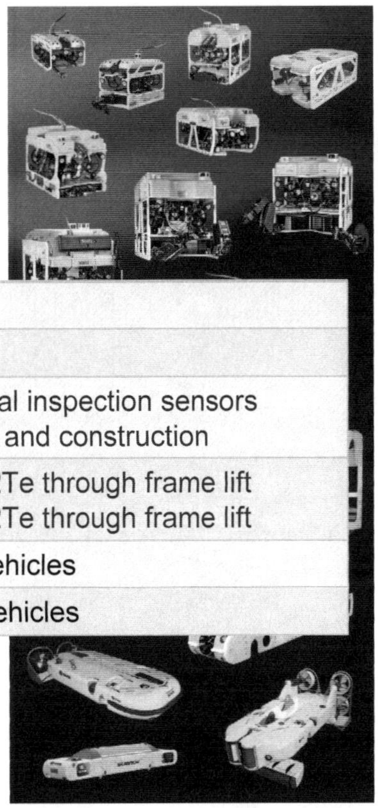

Class	Description
1	Observation ROVs
2A 2B	Observation with payload option – additional inspection sensors Observation - light intervention, survey and construction
3A 3B	Work class < 100kW < 200kg payload < 2Te through frame lift Work class > 100kW > 200kg payload > 2Te through frame lift
4	Towed & bottom crawling vehicles
5	Prototype or development vehicles

Contents

The problem with Class to define ROV systems

- Vehicle design and tooling solutions vary
- What is work ?
- Changing markets
- New technology and operational innovations

Contents

What is work ?

Credit - Fugro

Contents
What is work ?

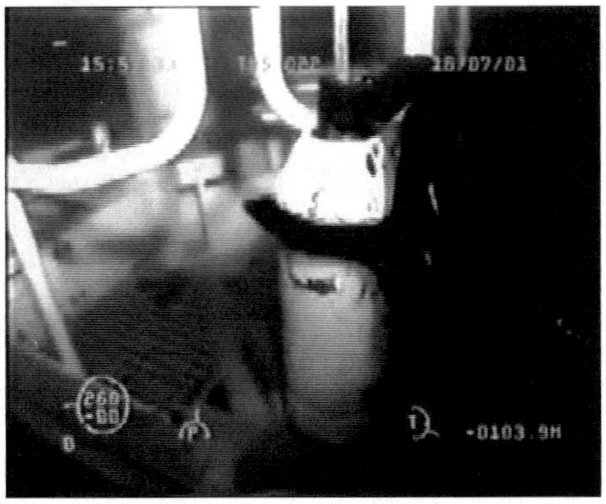

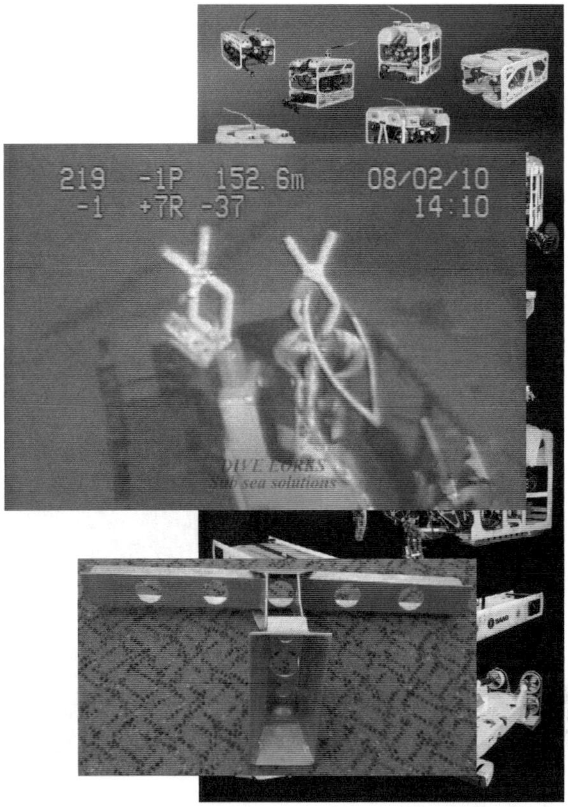

Credit – Fugro & DiveWorks

SAAB

Contents
Grout Hose Cutting tool

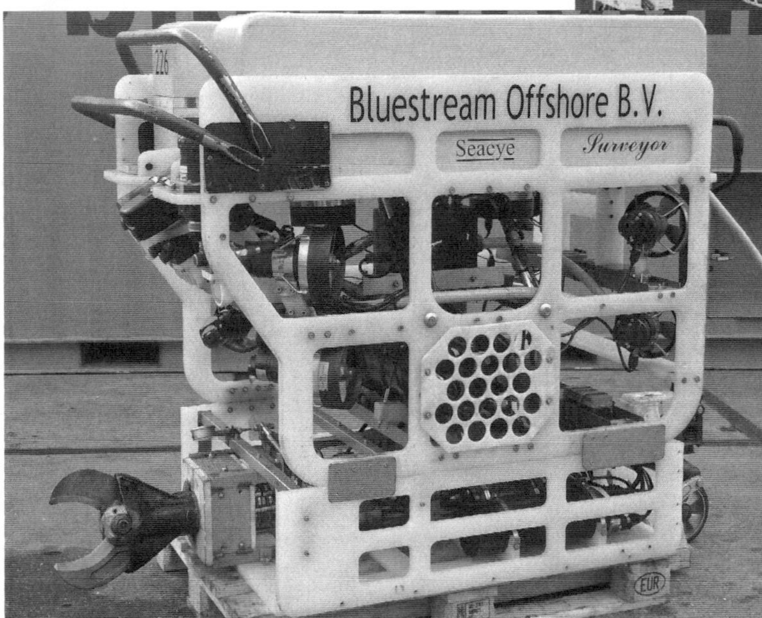

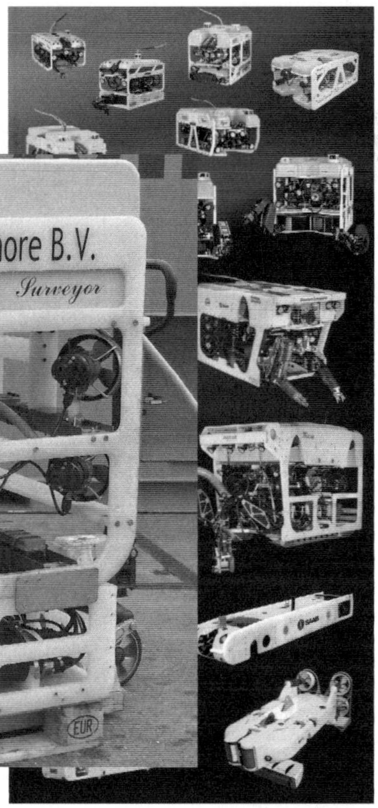

Credit – Bluestream

SAAB

Contents
Custom tool skids

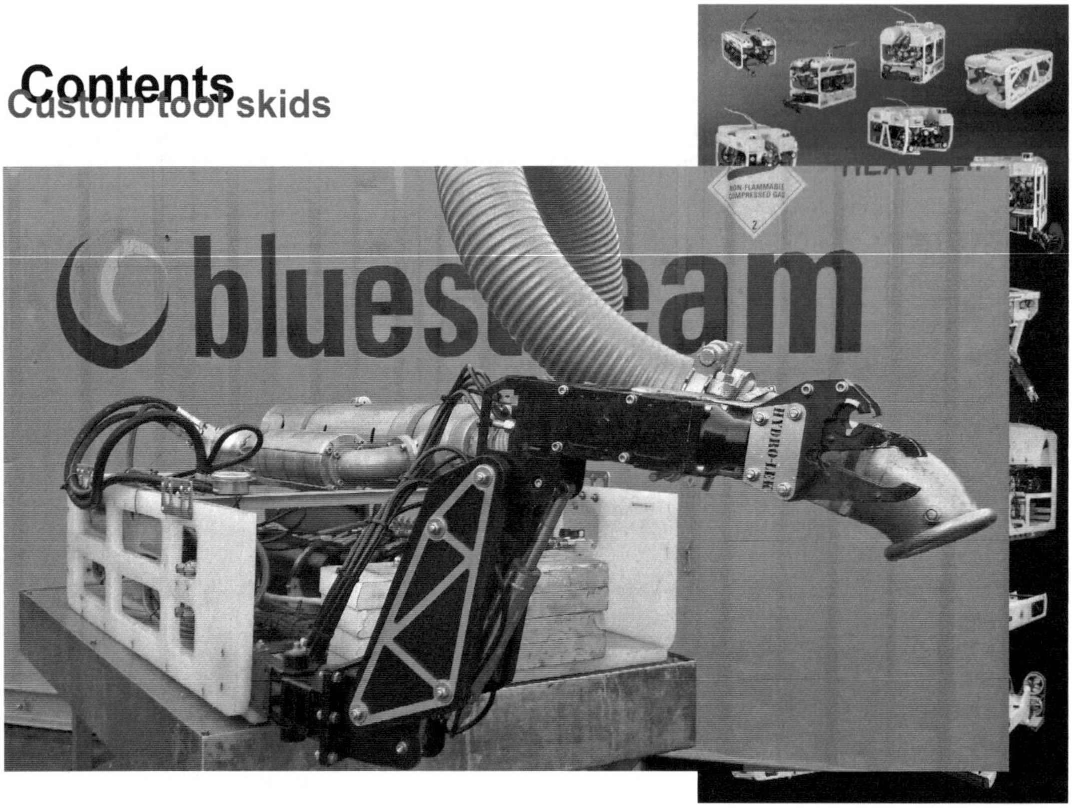

Credit – Bluestream

SAAB

Contents
Valve Operations

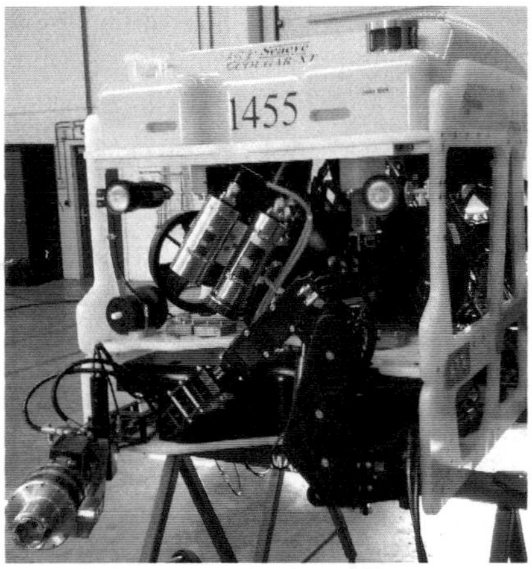

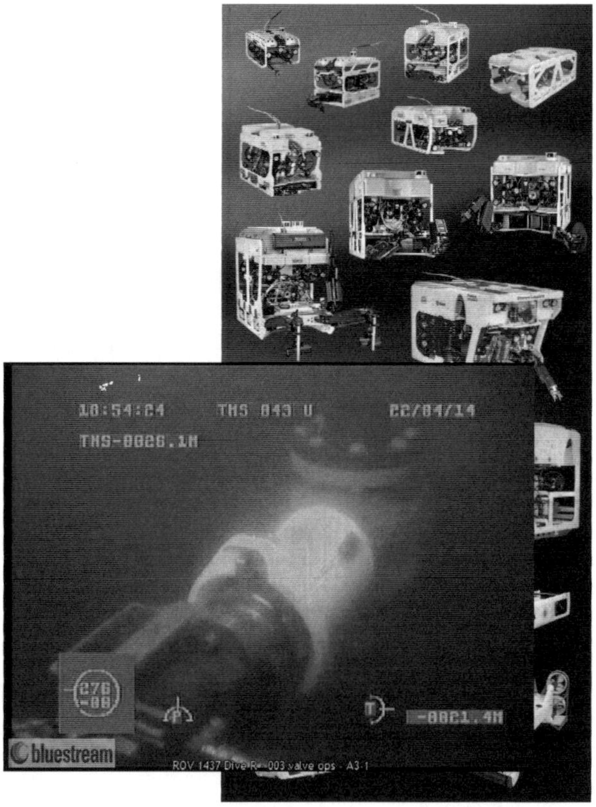

Credit – Bluestream

SAAB

Contents

Leopard conducting weld inspection

Credit – DiveWorks

SAAB

Contents

Structures deployed in Gullfaks field

Credit – Bluestream & Seaway Heavy Lifting

SAAB

Contents

Changing Markets

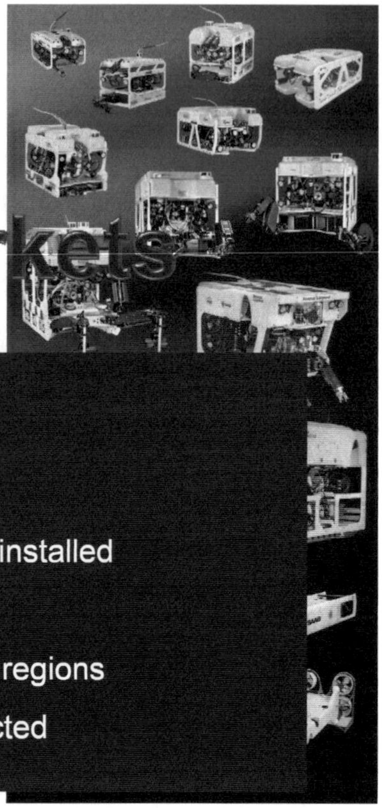

- BOP emergency override – API 53
- Field developments at greater depths
- Larger and more complex infrastructure installed
- Rapid growth in use of FPSO's
- Developments commenced in the Arctic regions
- Significant decommissioning work expected

SAAB

Contents

Renewables

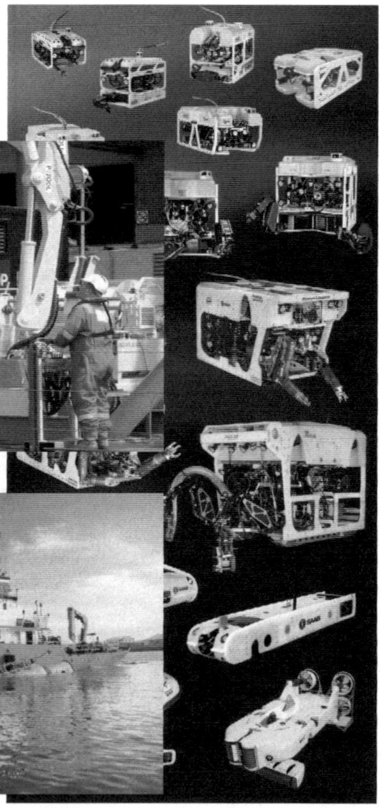

SAAB

Contents
Hydro Industry

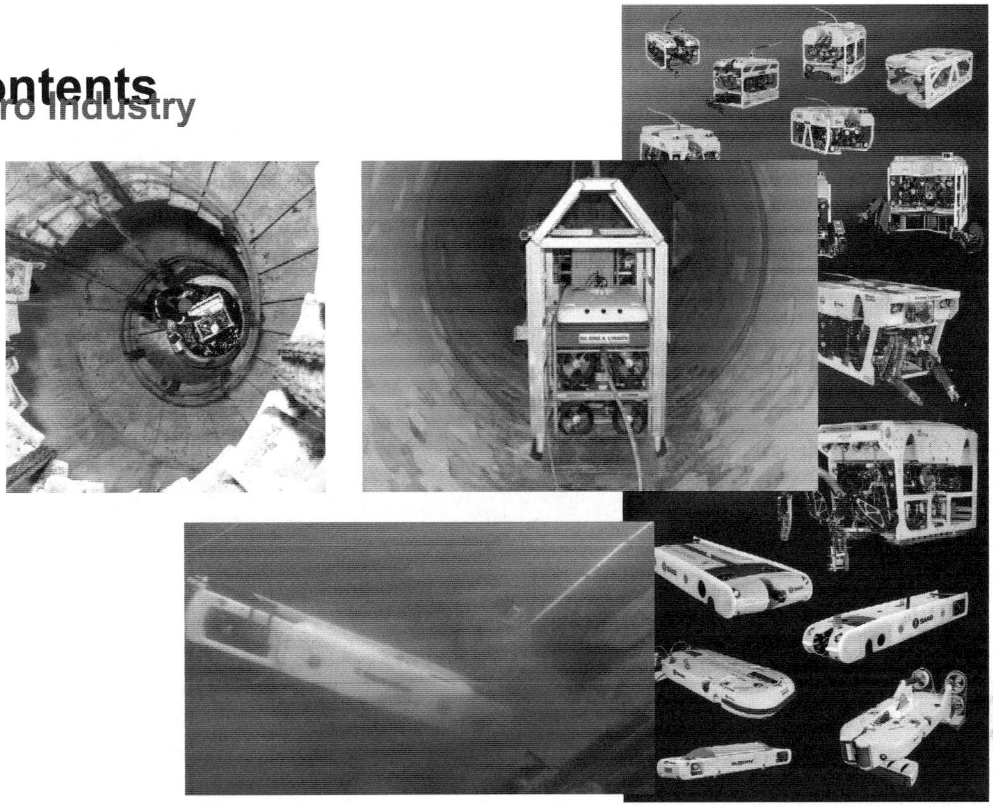

SAAB

Contents
Scientific and Aquaculture

SAAB

Contents
Military and Security

Contents
Operational locations

Contents

New Technology

New Operating Concepts

SAAB

Contents

AUVs & High speed survey_vehicles

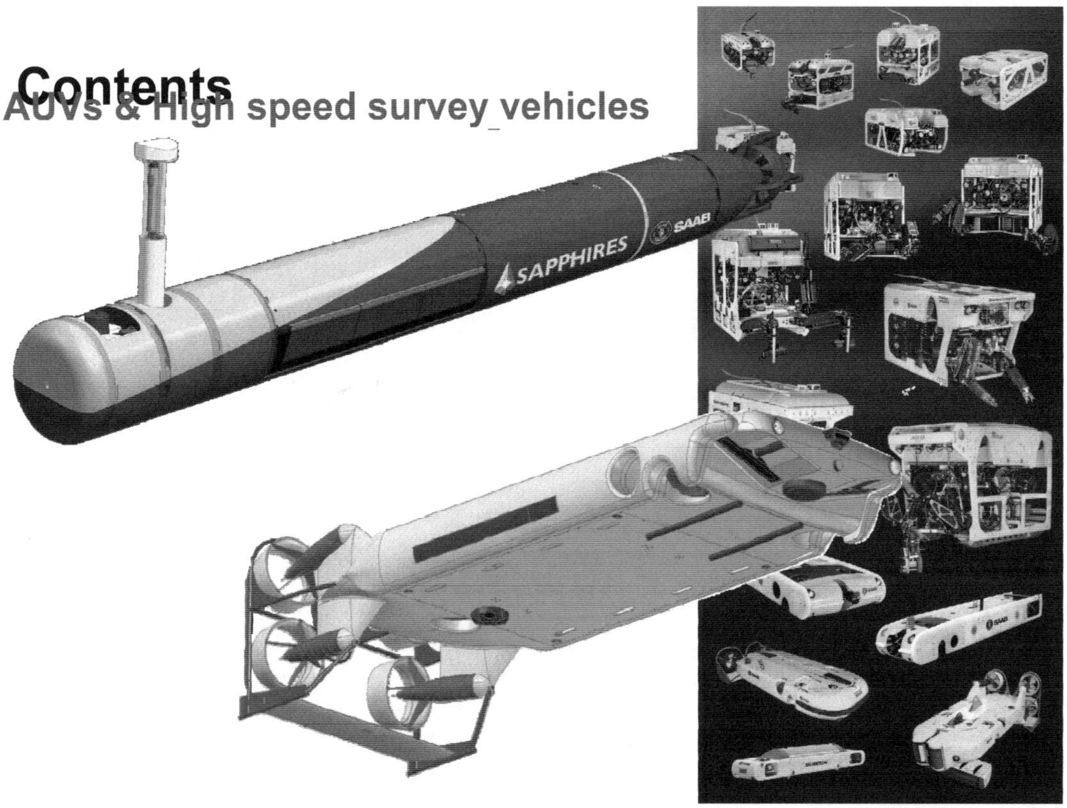

SAAB

Contents

Hybrid vehicles, electric torque tool & thr

SRV seabed infrastructure

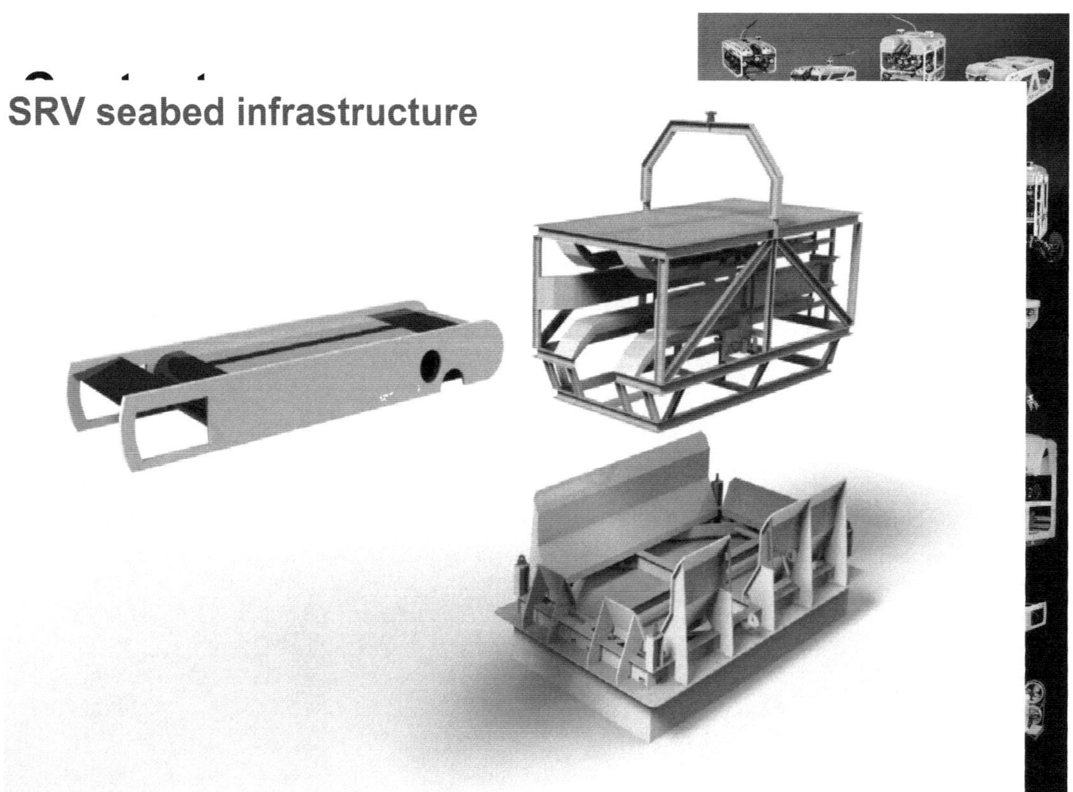

Contents

ROV / USV

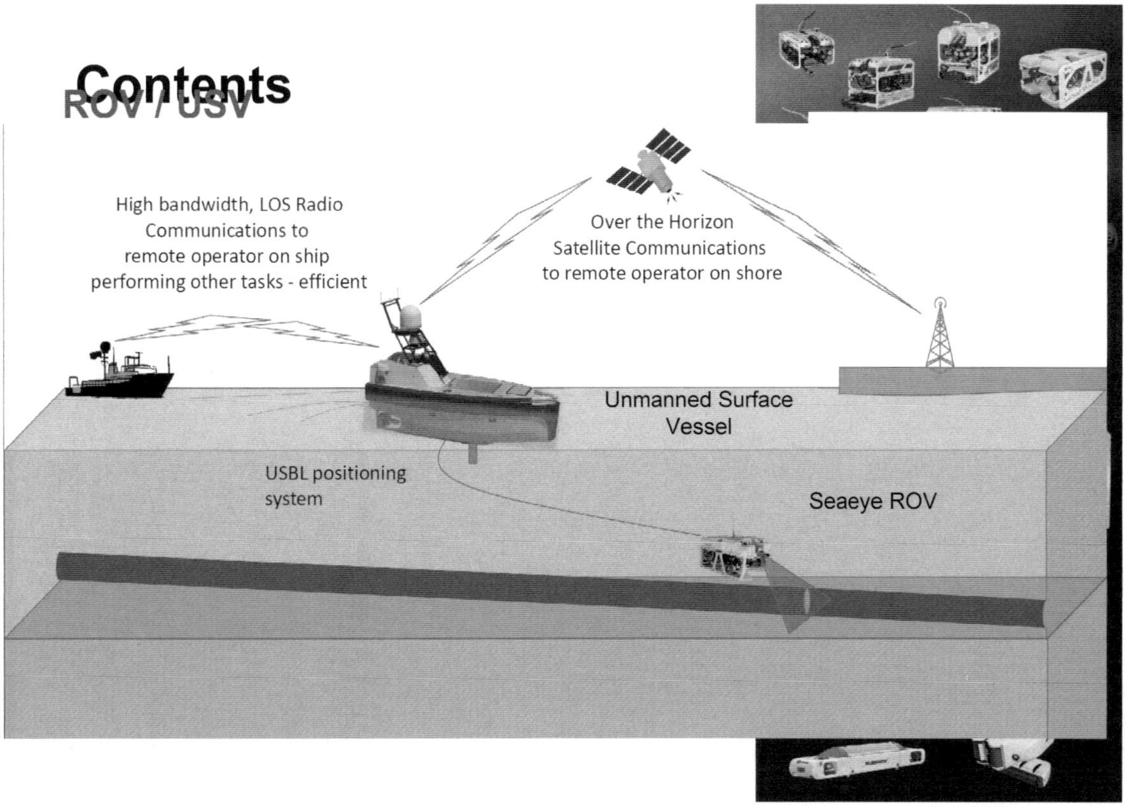

Contents

Vehicle Class -

Too many variables

Contents

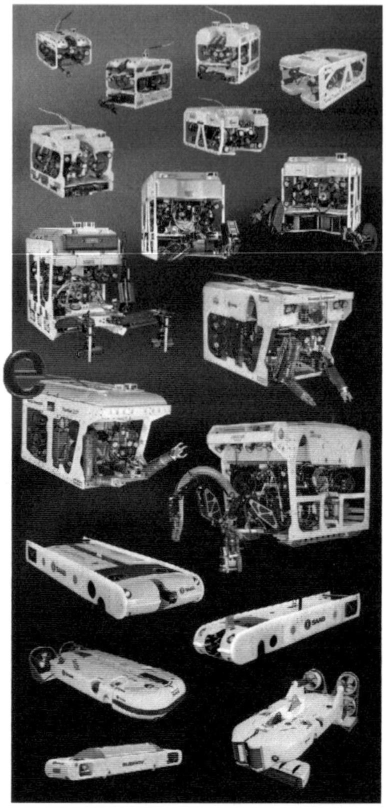

Vehicle Age

Contents

How old ?

Contents

How old ?

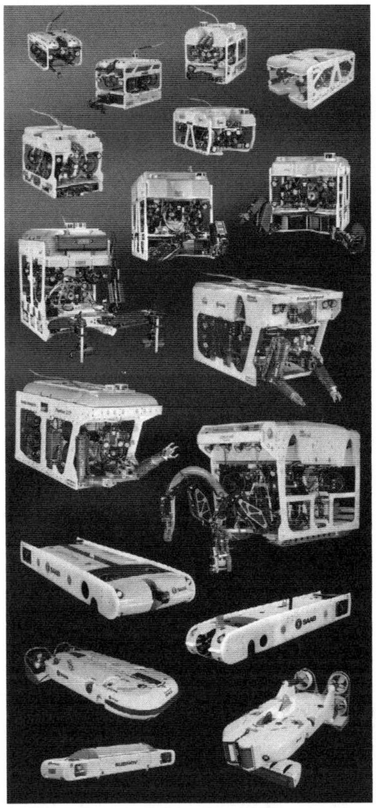

Contents

Summary

- Horse Power alone is not relevant
- Power costs
- Cost risk in over / under-specifying
- ROV classification not simple
- Age not very relevant

- Focus on what tasks and where

Contents

Forget HP, Age & Class

Think TASK

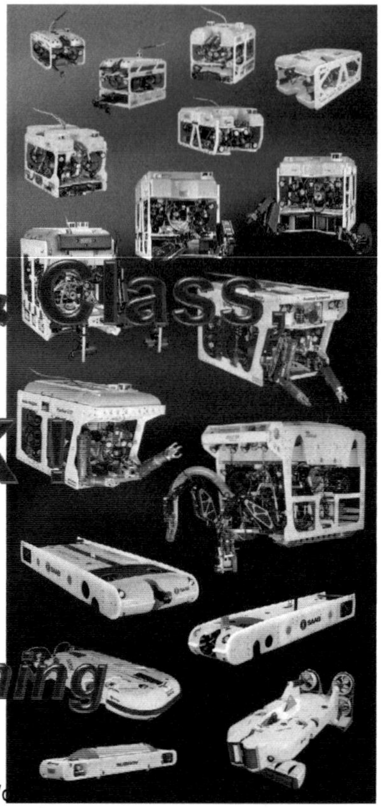

Thanks for listening

Thank you to Bluestream. Fugro and DiveW

for their assistance with operational material

SAAB

SAAB

QUESTIONS ?

www.seaeye.com

New ROV Technologies to Penetrate the USS Arizona

Evan Kovacs, Marine Imaging Technologies; Brett Seymour, National Park Service; William Lange, Advanced Imaging and Visualization WHOI

Introduction

The National Park Service (NPS) assumed management of the USS Arizona Memorial in 1980. The 608ft battleship resting in Pearl Harbor represents the nations premier Naval memorial honoring both the Dec 7, 1941 Japanese attack on the United States and the single greatest loss of life aboard a ship in US military history. The site is closed to all non-official, or recreational diving, and the NPS prohibits any diving within the ship out of respect for the nearly 1000 sailors and marines still entombed within the hull.

Since 2000, the Park Service has been studying the condition of the current submerged 100-year-old warship in an effort to address preservation options, oil release from interior bunkers, and any impacts to the Memorial structure that is suspended above the wreck – the site where nearly 1.8 million visitors pay their respects annually. Recognizing that limiting scientific study to the exterior of a fully submerged ship in the ocean environment, would mean that the NPS was unable to fulfill its mission to "preserve unimpaired natural and cultural resources…for the enjoyment, education, and inspirations of this and future generations" we concluded that some deployment of remotely operated technology inside the ship would be required.

After consultation with USS Arizona and other Pearl Harbor survivor groups, the NPS began to outline a scientific approach to inventory, sample and analyze the interior environment of Arizona in 2000 in an effort to understand the level of preservation within the ship. To assist with this complex research the NPS partnered with the Advanced Imaging and Visualization Laboratory (AIVL) at Woods Hole Oceanographic Institution (WHOI) and Marine Imaging Technologies

(MITech) to collectively identify and design ROV technology applications for deployment within the ship.

History

MITech and AIVL-WHOI have collaborated on the development of two new classes of ROVs for the scientific, inspection and documentary communities. The first class is a fly-away "cinematography class" ROV designed for close up inspection and survey tasks and obtaining ultra-high quality scientific imagery and cinematography quality underwater content not possible from industry and scientific ROVs. The first prototype ROV became operational in late 2015 and carries an immersive HD Panoramic imaging system and a suite of UHDTV Cinematography underwater camera systems. The second class of vehicles being developed by Mitech and AIVL-WHOI is the new Penetration Class ROV of which is based upon the need to complete long duration inspection of interior spaces in submerged objects and wreck sites not accessible by larger ROVs and Submersibles. The Penetration ROV is not battery limited and thus able to stay inside a shipwreck for several days at a time allowing longer term scientific environmental monitoring and survey tasks.

The 11th Hour is a specialized penetration remotely operated vehicle (ROV) built by MITech and WHOI's AIVL to access the interior environment of the submerged battleship USS Arizona in partnership with the NPS. The Park Service had utilized an ROV for interior survey previously in 2000 deploying a micro ROV built by VideoRay to visually assess the interior condition of the ship. In 2014 the NPS, AIVL and MITech began discussing the possibilities of a custom ROV that would take advantage of new technologies in high resolution digital imaging and to prototype a spooling tether management system allowing for longer, safer penetrations, scientific sampling and high resolution imagery for volumetric modeling. The design for the ROV was developed in early 2015 and the NPS co-funded some early development of the tether spooling technology. In December 2015 a test was conducted with an early prototype vehicle on the USS Arizona. After the test a 6 month design, engineering and testing timeline was developed for the next phase.

Due to the NPS's in ability to fund the entire project, potential sponsors and media partners were solicited. After substantial delays of external project funding, the build project was delayed until a mere 6 weeks before a scheduled expedition leading to a very compressed timeline for the ROV build and development. Due to the limited time constraints, the team worked with a existing ROV components for the internal ROV boards, propulsion and system control and spent the majority of the limited development and fabrication time on the tether management system, telemetry, power, scientific sampling, imaging design and systems integration on the vehicle.

Technology Development

<u>The Problem</u>

One of the major issues with internal shipwreck exploration with micro ROVs is the snagging of tether deep within a wreck and accumulating drag on the tether as the vehicle travels past successive bends in passageways and other interior structure. As the vehicle penetrates further into a wreck, there is a point it can no longer pull the weight or drag of the tether and that will end its forward progress. Additionally a snag at this could point could mean the loss of a vehicle if the vehicle cannot extricate itself. Even under normal circumstances, getting the vehicle in or out of a wreck generally requires a diver to be in voice communications with the ROV pilot and they work as a team pulling in or feeding out tether as the vehicle requires. This is not a feasible solution in deep-water environment when launching a micro ROV from a larger ROV or submersible.

<u>The Solution - Onboard Tether Management System</u>

The tether management system (TMS) for the 11th Hour ROV is a patent pending spooler on the back of the vehicle which allows the ROV to carry approximately 250' of 3/16" hybrid power and optical tether. The system does not require a slip ring and is purely mechanically driven by 2 motors making it unique in the micro ROV environment. The prototype performed reasonably well despite the difficult environment presented several decks down within USS Arizona but the TMS was also subject to jamming due to falling debris inside the ship and the open architecture of the ROV. These issues were largely the result of the extremely condensed timeline and have been subsequently modified with an updated redesign and gearing of the spooler.

Imaging Systems

The requirements for the project included dual 4K UHDTV imaging systems which were developed specifically for the 11th Hour and configured to provide 3D content. The dual 6G SDI camera platform was muxed with all vehicle and scientific sensor data on the one fiber in the tether. The high resolution cameras were recorded topside at varying resolutions for the assorted needs of the science and media partners and will ultimately be used to create volumetric visual reconstructions of the internal decks to help the NPS in their continued stewardship of the site. In addition to the 4K UHDTV 3D Stereoscopic system, there were numerous SD cameras for situational awareness, TMS monitoring and rear flying that were also muxed within the telemetry system.

Telemetry and Power

Although a VideoRay Pro 3 board set was used as the vehicle core, the 11th Hour runs off 300V and a single fiber optic communication to accommodate the very narrow diameter tether necessary for the TMS. The telemetry and power system are custom built to interface between a host of different data formats and power voltages depending on the application and vehicle. Converting the older VideoRay Pro 3 to fiber optics and 300v power worked flawlessly during interior investigations on USS Arizona and could potentially be the core of the next vehicle built.

Scientific Probes

During the USS Arizona mission in 2000, there was a desire to capture unadulterated physical samples and monitor real time environmental water chemistry as the ROV investigated the interior lower decks on the ship. A reliable sampling device was never achieved and interior water chemistry could not be monitored in real time, only downloaded from a YSI environment sonde mounted to the vehicle. To accomplish NPS research needs, the team built a custom sealed sampler for water and sediment sampling and also carried a YSI environmental probe that tied into the telemetry system that allowed real-time and recorded environmental data such as dissolved oxygen, pH, salinity, temperature, and oxygen reduction potential – all of which are important for the NPS' current research into the lifespan of the battleship submerged in Pearl Harbor for 75 years.

11th Hour USS Arizona Deployments

In December 2015 an early prototype of the vehicle was tested to evaluate the system power needs and telemetry integration along with several gearing and tension designs on the tether spooler. The deployments led to refinements in spooling speeds, gearing and most importantly, tensioning of the tether while spooling in and out. HD3D Stereoscopic camera systems were integrated for image testing and interior penetration into the ship was limited to the upper decks where an emergency recovery was possible if necessary.

In August 2016 the NPS, MITECH and AIVL-WHOI collaborated with PBS and Curiosity Stream to conduct an interior survey and collect samples from USS Arizona's third deck. The expedition would also produce a scientific documentary showcasing the NPS and partners stewardship of USS Arizona to commemorate the 75th anniversary of the ships sinking in 1941. Unfortunately due to an extremely compressed engineering and fabrication timeline, the independent components of 11th Hour were fabricated in Massachusetts but assembled for the first time in Hawaii. As with any prototype engineering effort, modifications were required to make the ROV functional and extensive on-site fabrication of components was conducted in the garage of a rented house. After a week of 20-hour days, the 11th Hour was assembled and tested in water. Although the tether management system worked as designed, wiring harness issues necessitated a rebuild. After a few days of modifications, re-engineering of components and much soldering, the 11th Hour was tested on Arizona's 2nd deck and planning for a 3rd deck investigation was undertaken.

The 11th Hour was larger and heavier than originally designed which required a bit of pilot adjustment once inside the wreck. A battleship by nature is designed for war, not comfort. The hallways were small and tight and the ROV could barely turn around due to the sized of the vehicle and tether spooler. With the addition of the YSI probe mounted on the ROV, this became even more difficult. The typical penetration plan involved traditional feeding of tether from a support diver positioned at the point of entry while navigating down hallways to a predetermined area of the wreck. From there, the onboard tether spooler was utilized while entering a cabin, effectively laying tether upon entry, and then backing out of the cabin while retrieving the tether back on to the spooler. This proved to be a highly effective method for systematic penetrations and imaging.

With the success and reliability of the tether management system of the ROV on the 3rd deck, the tether support divers were recalled from the ROV access points and the vehicle operated autonomously inside the wreck. For the next full day of operations the 11th Hour worked inside the wreck utilizing the tether management system with no external tether assistance and surveyed all open officers staterooms in the aft section of Arizona's 3rd deck. Some of the staterooms had been surveyed during the initial 2000 interior investigation, but many were inventoried for the first time and interestingly, the accumulation of sediment was noticeably different in the rooms visited before – suggesting that the internal spaces of Arizona were subject to changes contrary to what we had expected. The most striking of all the discoveries was the discovery of organic materials, including clothing and uniforms hanging in several of the officer's staterooms and the imaging of the various microbiological colonies – most notably on the ceilings. These discoveries all aid the NPS to understand the current environmental conditions within the ship and attest to the remarkable level of preservation deep within Arizona's hull.

The August 2015 trip effectively ended when microbiological formations (stalactites) dislodged from the overheads and landed on the vehicle causing jamming and loss of the vehicle thrusters. The vehicle was retrieved from the 3rd deck using the onboard tether management system. The non-traditional recovery ended up destroying much of the internal gearing that was built by hand during the Hawaii build. Despite the operational crippling of the vehicle, the team ultimately inventoried most of the accessible staterooms, cabins and hallways in Arizonas 3rd deck and captured the most moving imagery and comprehensive environmental data to date inside the wreck.

After the August mission the 11th Hour was internally rebuilt with new motors and gearing for a December 2016 USS Arizona interior mission solely dedicated to science with successful collection of water, sediment and microbiological samples. In addition, real time data from the multiparameter YSI environmental probe was logged and cross-referenced with continuous video time code and mission logs providing detailed location of data. Dissolved oxygen, pH, salinity, temperature and depth were among the most relevant data sets captured with direct bearing on the interior preservation of Arizona.

Conclusion

As the NPS continues its interior investigation and sampling within USS Arizona, Marine Imaging Technologies and Woods Hole Oceanographic Institution's Advanced Imaging and Visualization Laboratory will continue to refine and advance the 11th Hour to meet the mission requirements. To date with multiple deployment and missions depending on the 11th Hour's innovative tether management system, the vehicle has proven longer and deeper penetrations are feasible with less chance of major snags than previously thought possible. This new category of micro ROV with a custom tether management system, advanced imaging systems and custom scientific sensors and sampling systems, creates new possibilities for hazardous interior structural penetrations and can be scaled and adapted for a variety of operational depth ratings.

Pipe Crossing Protection of Subsea Pipelines

Benton F. Baugh, Ph.D., P.E.[1], Joe Pasztor[2], and Paul VanderLinden[3]

Abstract— A discussion of various methods or providing for the physical and electrical separation of subsea pipelines which cross on the ocean floor, including a movable pipe crossing, concrete revetment mats, and a concrete bridge.

I. INTRODUCTION

When subsea pipelines or umbilicals cross, special provisions must be made such that the pipelines and their coatings are not damaged due to direct electrical contact or abrasion due to sliding. Alternative methods of providing the necessary protection are discussed. It is critical that the pipelines remain separate and electrically insulated from each other. Basic methods discussed to accomplish this are a movable pipeline tee, concrete revetment mattresses, and a concrete bridge.

II. GLOSSARY

Revetment mattress: An assembly of concrete blocks interconnected by steel cables for laying over and stabilizing a pipeline.

Buckle arrestor: A short enlargement which limits the extend of the damage to a subsea when a buckling situation starts.

ROV: Remotely operated vehicle used to do subsea tasks.

III. DISCUSSION

Fig. 1 shows a first method which is a movable pipe crossing which contacts the pipe only with insulating roller type wheels. This style is pivoted at the center to match the angle of the pipelines and is also insulated at this point, giving three insulating areas between the pipelines. There is an upper half and a lower half which are sized to match the pipeline size to be engaged.

Fig. 2 shows and ROV or diver handle which releases the lowering slings used in installations when both pipelines have already been laid. This handle is shown at the upper right portion of the upper half as a round bar projecting inwardly

Fig. 3 is a top view of a movable pipe crossing, showing side rollers which deflect outwardly automatically when approaching the pipeline and then snap back in place to lock the movable pipe crossing to the pipe.

[1]Benton F. Baugh, Ph.D., P.E. is a principal of Baugh Consulting Engineers, Inc. and adjunct professor at the University of Houston, bbaugh@baughengrs.com.

[2]Joe Pasztor is a principal at Pasztor Machine Services, pasztormachinese@aol.com

[3]Paul VanderLinden, paulgvanderlinden@gmail.com

Fig. 4 starts a brief sequence on how the movable pipe crossing is installed with a new crossing pipeline. A buckle arrestor is located in the pipeline so it will touch down shortly before the crossing pipeline lands on the existing pipeline.

At the surface the movable pipe crossing is clipped to the pipeline and it will automatically hang below the pipelines as shown. How it is clipped to the pipeline at the surface will depend on the type pipelay operations and which vessel is being used. The angle between the pipelines will be known, so the unit is safety wired to that angle so that it easily lands on the lower pipeline at touchdown.

Fig. 5 shows that as the landing pipeline nears the existing pipeline, the ROV pushes the movable pipe crossing forward to alignment with the existing pipeline with one of its manipulators.

Fig. 6 shows that the pipeline continues to be laid down and the movable pipe crossing locks onto the existing pipeline.

Fig. 7 shows that the ROV has completed its task and the pipeline continues to be laid so the ROV is not shown. As a practical matter, the ROV is likely to remain in position as it is likely watching the touchdown point as the pipeline is laid. This means that there is functionally no added operational costs associated with this type installation of the movable crossing as this ROV is typically already there.

Figs. 8 – 13 is a sequence of operations for installing the movable pipe cross on pipelines which have already been laid. This is necessary when more conventional methods of separation have failed due to currents, bottom scouring or pipe shifting on the seafloor due to storms or anchor drags. This method shows that the upper pipeline has been picked up and the movable pipe crossing is lowered by a cable and a sling set. Both these operations can likely be accomplished by a single ROV service boat. Mats will likely be landed, the pipeline picked up and landed on them, but this requires the ability for some lateral movement of the upper pipeline. Otherwise, two working cables would be required from the vessel.

Fig. 9 shows that the movable pipe crossing is lowered onto and automatically clips or locks to the upper pipeline. A diver or an ROV can move the movable pipe crossing into position for this engagement.

Fig. 10 shows that the supporting cable is lowered and the unit flips upside down. In this motion the sling set automatically disengages from one side. The ROV or diver can disengage the other side of the sling set now or as the final step by simply pushing the handle as seen in Fig. 3.

Fig. 11 shows that the movable pipe crossing is moved along the upper pipeline into alignment with the lower pipeline by the diver or ROV.

Fig. 12 shows that the upper pipelines now lowered onto the lower pipeline. Again, it is presumed that the lower half is safety wired into proper alignment with the upper half for ease of operations.

Fig. 13 shows that the slings are now removed and the job is complete. The two pipelines are triple insulated from each other, and if the pipelines shift now, the movable pipe crossing simply moves with them. It is a permanent solution even if the pipelines move.

Fig. 14 shows a more conventional method which is to separate the pipelines with a passive material such as sand bags or concrete revetments mattresses. A concrete revetment mattress is shown laid across a first or the existing pipeline. This not only provides the potential for separating this pipeline from and upper pipeline, but also provides weight to tend to keep the pipeline in place.

Fig. 15 that the upper pipeline has been laid onto the first or lower revetment mattress covering the lower pipeline and a second revetment mattress has been laid onto the top of the upper pipeline.

Fig. 16 shows a drawing of a revetment mattress such as shown in Figures 14 and 15, and a smaller drawing showing that the concrete blocks are twelve inches square by about six inches thick.

Fig. 17 shows a photograph of an actual concrete revetment mattress laid over a pipeline on land as a demonstration.

Fig. 18 shows a truckload of revetment mattresses to give a better idea of the sizes.

Figures 19 and 20 show frames prepared to lower offshore revetment mattresses into position on subsea pipelines.

Fig. 21 shows a revetment mattress which is touching down on the seafloor during the installation procedures.

The vessels making the trip to install the concrete revetment mattresses need to have the capacity to not only lower the mattresses and release from them when they are landed, but also to orient them when they are in the correct location. If the mattresses are to be reinstalled after the pipelines have been moved by currents or shifting seafloors, the vessel(s) must be able to not only pick up the upper pipeline, but also to move it laterally over the new mattress location.

Fig. 22 shows a third method which is to install a precast concrete bridge over the lower pipeline. This method would also require a lift vessel(s) capable of handling the weight of this bridge and orient it. Also in the case of reinstallations, the vessel(s) would be required to move the pipeline laterally to get it onto the bridge.

Each of these methods can provide adequate separation of the pipelines and insulation between them. Each have their own strengths and difficulties depending on the nature of each job.

Fig. 1 Movable pipe crossing separating two pipelines

Fig. 2 Showing ROV release for sling set at the upper right

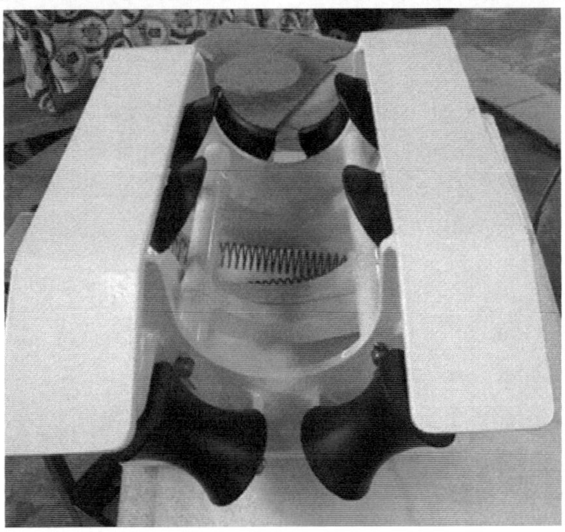

Fig.3 Bottom view showing rollers which lock onto the pipeline

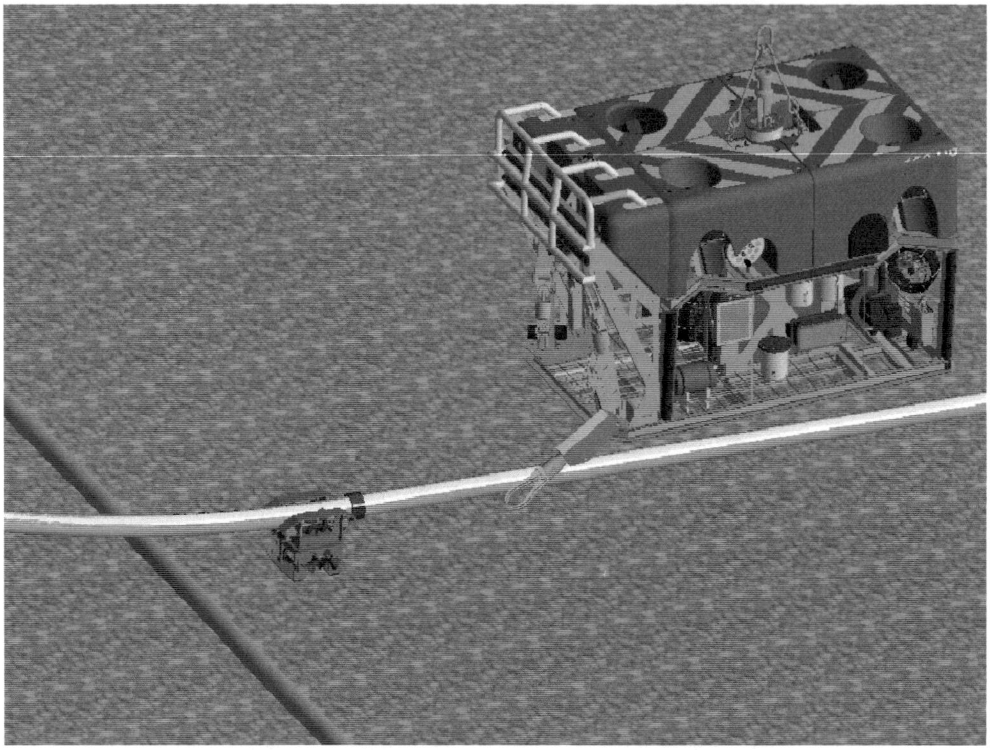

Fig. 4 Movable pipe crossing against a buckle arrestor in an original installation

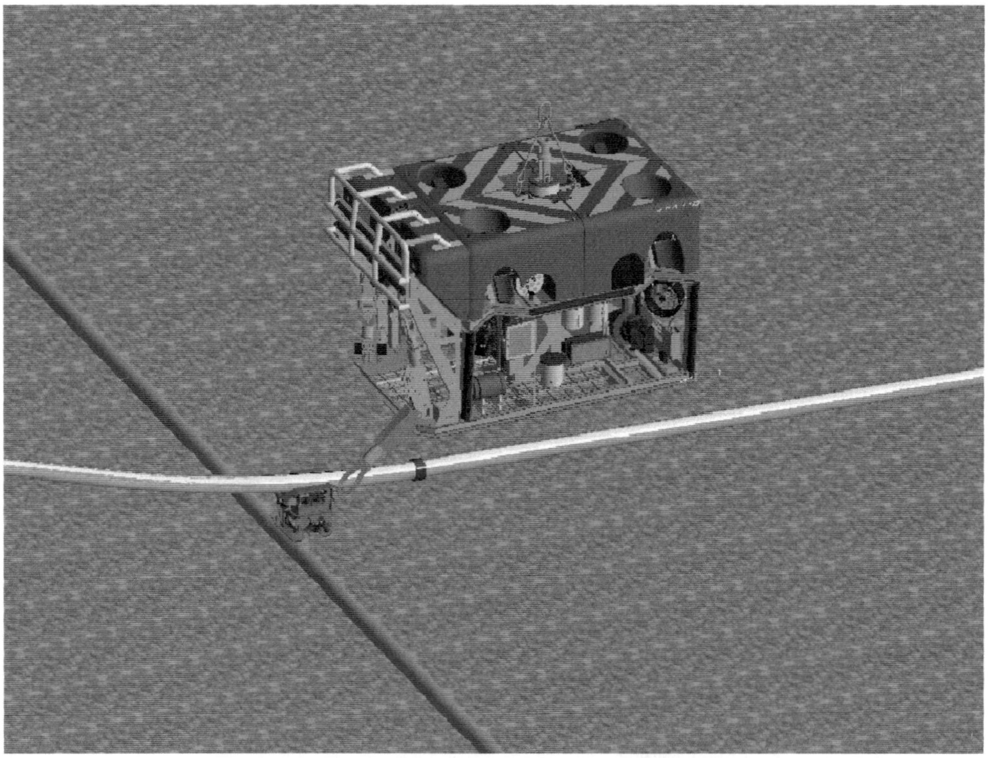

Fig. 5 ROV moving the movable pipe crossing to alignment with the lower pipeline

Fig. 6 Upper pipeline lowered onto lower (existing) pipeline

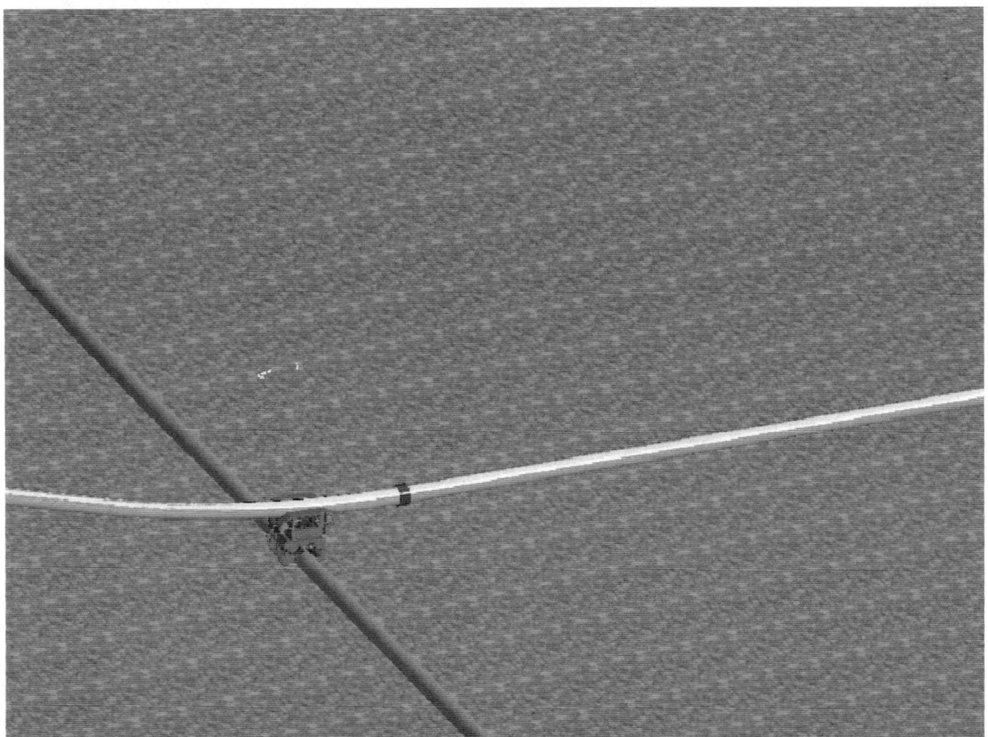

Fig. 7 Installation complete, pipeline laying continues

Fig. 8 Approaching upper pipeline

Fig. 9 Landed on upper pipeline

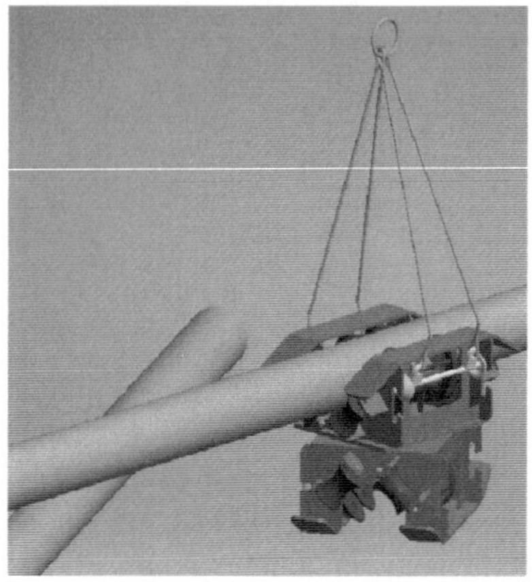

Fig. 10 Flipped

Fig. 11 Upper pipeline raised and Approaching lower pipeline

Fig. 12 Landed on lower pipeline

Fig. 13 Slings removed, fully installed

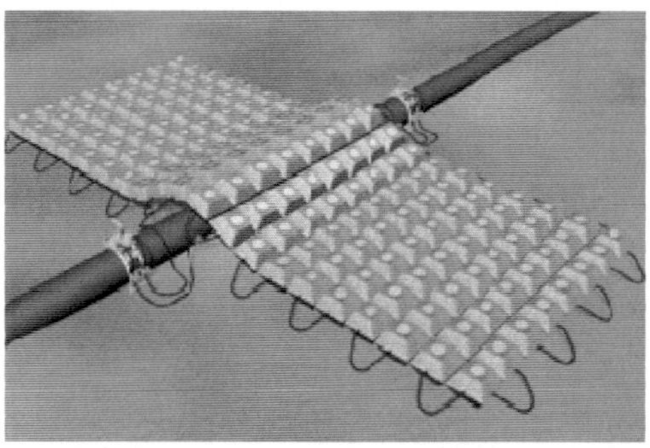

Fig. 14 Revetment mattress on pipeline

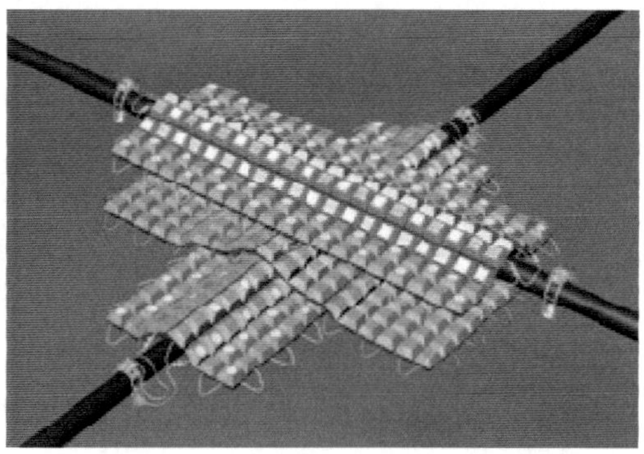

Fig. 15 Revetment mattress on crossing pipeline

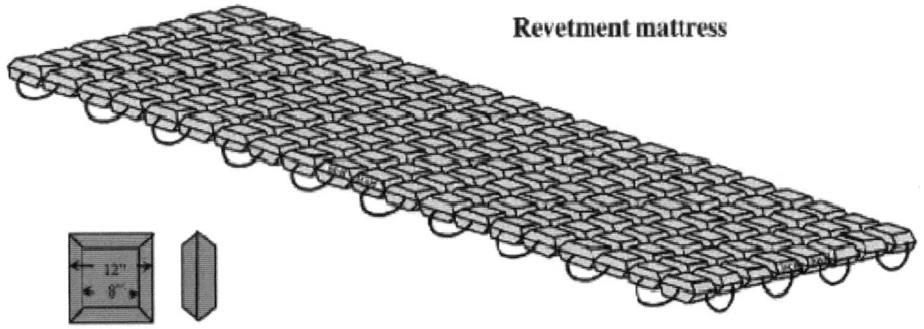

Fig. 16 Drawing of a concrete revetment mattress

Fig. 17 Photo of concrete revetment mattress on pipeline

Fig. 18 Truckload of concrete revetment mattresses

Fig. 19 Frame lowering a concrete revetment mattress

Fig. 20 Frame lowering a concrete revetment mattress

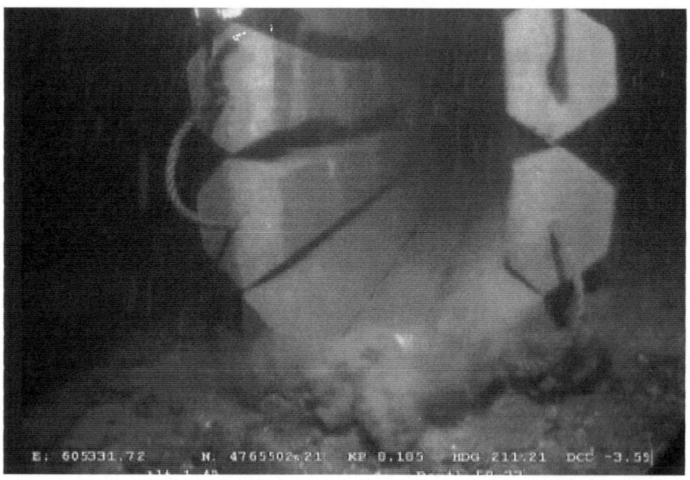

Fig. 21 Concrete revetment mattress touching down when landing

Fig. 22 Concrete bridge type pipe crossing

Vision based Assistance for Underwater Manipulations

Gun Rae Cho[1], Sung-Mun Hong[1], and Ji-Hong Li[1]

Abstract—**In the paper, a vision based assistant technique for underwater manipulations is proposed. Especially, 3D target position estimation issue and inverse kinematics issue are addressed. Regarding target position estimation issue, a technique which can estimate 3D position of the objects utilizing 2D position information on the two images from cameras deployed in stereo is addressed. Via touch screen, the 2D position information is fed by operators for guaranteeing reliability of the estimation results, which is one of the most important points in the underwater interventions. Regarding kinematics issues, a kinematics solution based on the weighted damped least squares is investigated to overcome the limitations: workspace limit and singularity, unintended changes of the robot configurations, and joint position limits. By designing appropriate cost functions for each limitation, the proposed kinematic solution provides a solution optimized in the manner of weighted norm, even when the manipulators meet those limitations. Through the laboratory experiments, the performance of the proposed assistant technique is verified.**

I. Introduction

Underwater robots have been used widely for various application fields: e.g., science, fisheries and aquaculture, military, security and safety, oil industry and constructions [5][19]. Especially, the underwater intervention tasks are mainly performed by using manipulators [19]: the tasks involve cable cutting, unshackling, opening and closing covers, and installing instruments. These tasks are commonly achieved by teleoperation: the operator commands every motion of manipulators, using limited information of underwater situation via cameras and sensors [6][19][20]. The operators have to recognize spatial situation with the limited information and to tele-operate every degree-of-freedom (DOF) of the manipulator using input devices; thus, the process is time consuming and requiring well-experienced operators, resulting in high cost to perform the tasks.

To overcome the inefficiency in underwater manipulation, several research works utilizing autonomy have been reported. Most of them are focused on how to obtain the position of the object automatically in underwater situations: e.g., sonar sensor based technique [7], and the vision based approaches [8][9].

Although providing comfort to the operators, autonomy may not be well-accepted because that high reliability is demanded to perform intervention in underwater situation. The authors have interviewed with a person who had more than 10-year-experiences as a ROV operator in Oceaneering: According to his explanation, there are many auto-functions in ROV, but the operators seldom use them to prevent any risk of malfunction. Although being reasonable approaches in undetermined situation, moreover, the sensor based recognitions may cause problems in the underwater manipulations: i.e., inverse kinematics problem. It is because that the target positions obtained from sensors do not always satisfy the limitations of the manipulators. The limitations involve the followings: workspace limitation and kinematic singularity, unintended change in robot configuration, and joint position limits. For stable operations, therefore, it is required the kinematics algorithm which can provides appropriate results even when the target position is given as out of the limitations.

For the kinematics issue, the numerical approaches based on the Jacobian provide effective solutions even when the manipulators have complex geometries or kinematic redundancy [10][11]. The damped least squares (DLS) [12][13] provides simple way to overcome the singularity problems and have been verified with various applications: e.g., industrial robots [14] and redundant manipulators [15]. As another approach, the weighted least squares (WLS) has been applied to optimize the kinematic solution with physical meaning: e.g., energy minimization [16], torque minimization [17], and joint position limit avoidance [18]. In [18], in more detail, it was shown that the WLS is more effective to avoid the joint limitation than the approaches using kinematical null space design.

In the paper, a vision based assistant technique for underwater manipulations is proposed. The authors focus on '*assistance*' instead of '*autonomy*' because that the assistance function can be intervened by the operator and is easy to maintain the reliability to perform tasks. The paper focuses on the procedure operating manipulators to approach the objects. By only indicating target positions on the camera images via touch screen, one can make the manipulators approach to the target. The main issues regarding proposed assistant technique can be categorized as follows:

- Target position estimation using camera and touch screen,
- Inverse kinematics providing reasonable solution even when the target position exceeds the limitation of manipulators.

Regarding the target position estimation issue, a 3D position estimation technique utilizing two of 2D position from the

This research was supported by the project titled 'R&D center for underwater construction robotics', funded by the Ministry of Oceans and Fisheries (MOF) and Korea Institute of Marine Science & Technology Promotion (KIMST), Korea (PJT200539).

[1]G. R. Cho, S. -M. Hong, and J. -H. Li are with Marine Robotics R&D Division, Korea Institute of Robot and Convergence, 37666, Jigok-ro 39, Nam-gu, Pohang, Gyeongbuk, Ref. of Korea {sandman, sungmun9, jhli5}@kiro.re.kr

cameras deployed in stereo is proposed. The 2D positions on the camera images are fed by the operators via touch screen. The touching process can reduce burden to tele-operate the manipulators; moreover, it provides easy way to guarantee the reliability of target position estimation results by utilizing the position information indicated by the operators.

Regarding the inverse kinematics issue, a solution based on the weighted damped least squares (WDLS) is provided for underwater manipulations. When deriving the solution, it is considered the practical issues that the target position exceeds the kinematic limitations:

- When the target position is out of the workspace,
- When the manipulator passes singular positions and has unwanted configuration change near singularity, and
- When joint position limits are violated.

By utilizing WDLS [10][11], aforementioned limitations are resolved: the workspace and singularity issues are handled with the DLS approaches; and, the joint limit issues are resolved using the WLS scheme.

The performance of the proposed method is verified through the laboratory experiments with webcams and an industrial manipulator.

II. Target Position Estimation via Touch Screen

Fig. 1 show the schematic diagram of vision system. The system has two 2D cameras deployed in stereo and the robot manipulator. Besides, there is a touch screen with which the operator can monitor the images from the two cameras and can indicate the position of the target on each image. The goal of the target position estimation is to estimated 3D position of the target utilizing the touch screen input from operators.

From Fig. 1, one can find the coordinate systems for the target position estimation: W denotes the world coordinate; *Ci* the focus coordinates of each camera, respectively; and, *Ii* the image coordinates of each camera, respectively; and $i=1,2$ denotes the camera number. The relative positions between W and *Ii* are given by the geometrical relationship between the manipulator and the cameras; whereas, the positions of *Ci* are decided by the positions of *Ii* and the focal lengths, *f*, of each cameras. In the experiment in Section VI, we used the camera, HD-3000 model from Microsoft which has the focal length $f_x = 617.98mm$, $f_y = 625.32mm$.

The proposed algorithm to estimate the target position can be categorized into the three steps as follows:

- Touch screen input treatment to obtain the target positions w.r.t the camera focus coordinates,
- Generation of the two lines connecting each camera focus and the target position, and,
- Estimation of 3D positon of the target by selecting the nearest point between the two lines.

Touch screen input treatment to obtain the target positions w.r.t the camera focus coordinates: When the operator indicates the target position on each camera image via touch screen, one can determine the 3D positions of target, *Pi*, w.r.t. the camera focus coordinate, *Ci*, by choosing the distance along z-axis arbitrarily. If one chooses the distance as z=1, the positon is given as follows:

$$ {}^{C_i}P_i = \begin{bmatrix} x_i/f_x & y_i/f_y & 1 \end{bmatrix}^T \tag{1}$$

where, x_i and y_i denotes the 2D positions of the target on the camera images, which are indicated by the operator via touch screen; and, $i=1,2$ denotes the camera number.

Generation of the two lines connecting each camera focuses and the target position: By using the position information in eq. (1), one can generate lines which connect camera focus points and the target position in (1).

First, the positions of camera focus points are given by the geometric information and the focal length of the cameras, and can be described w.r.t. the world coordinate as follows [2]:

$$ {}^{W}P_{C_i} \equiv \begin{bmatrix} c_{xi} & c_{yi} & c_{zi} \end{bmatrix}^T \tag{2}$$

Besides, the target positions in (1) can also be described w.r.t. the world coordinate as follows [2][3]:

$$ {}^{W}P_i \equiv \begin{bmatrix} p_{xi} & p_{yi} & p_{zi} \end{bmatrix}^T = {}^{W}_{C_i}R\,{}^{C_i}P_i + {}^{W}P_{C_i} \tag{3}$$

where, ${}^{W}_{C_i}R$ denotes the rotation matrix of the camera focus coordinate, *Ci*, w.r.t. the world coordinate, *W*. One can obtain the equation of the lines connecting the points in (2) and (3): when z is regarded as independent variable, the line equation is obtained as follows:

$$ \overline{{}^{W}P_i {}^{W}P_{C_i}} \equiv \begin{bmatrix} \dfrac{p_{xi}-c_{xi}}{p_{zi}-c_{zi}}\left(z-c_{zi}\right)+c_{xi} \\ \dfrac{p_{yi}-c_{yi}}{p_{zi}-c_{zi}}\left(z-c_{zi}\right)+c_{yi} \\ z \end{bmatrix} \tag{4}$$

where, $i=1,2$ denotes the camera number.

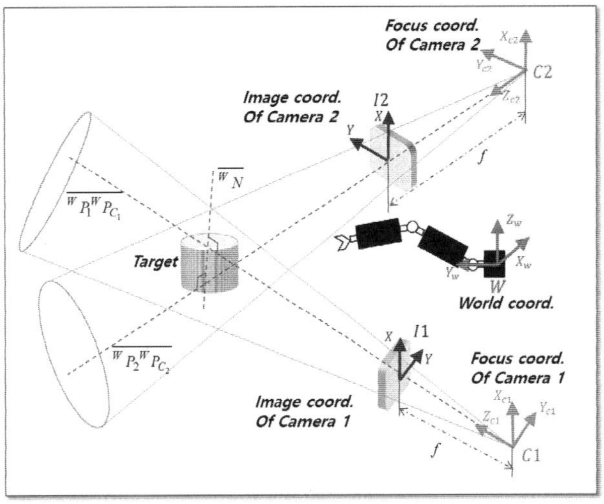

Fig. 1 Schematic diagram of vision system

Estimation of 3D positon of the target by selecting the nearest point between the two lines: By obtaining the nearest point of the two lines in (4), one can fix the 3D position of the target: i) first, the normal vector perpendicular to the two lines in (4) is obtained, ii) second, the intersection points between the normal vector and each line in (4) are obtained, iii) by averaging the intersection points, the target position is estimated.

On can obtain the normal vector by performing cross product as follows [4]:

$$^{W}N \equiv \begin{bmatrix} n_x & n_y & n_z \end{bmatrix}^T = \left({}^{W}P_1 - {}^{W}P_{C_1}\right) \times \left({}^{W}P_2 - {}^{W}P_{C_2}\right) \quad (5)$$

The vector in (5) denotes the direction which is perpendicular to both of the two lines in (4); one can describe the points on the normal vector in (5) as follows:

$$\overline{^{W}N} = \begin{bmatrix} \frac{n_x}{n_z} z + \alpha & \frac{n_y}{n_z} z + \beta & z \end{bmatrix}^T \quad (6)$$

where, α, β denotes the offsets along x and y axis, respectively. Note that the offsets are introduced in (6) to describe parallel transference of the normal vector in (5); the offsets in x, y axis are considered because that the direction of the normal vector is given as constraint, and it is enough to consider offsets in the two DOF instead of three DOF for special transference.

From (5) and (6), one can describe the constraints of the intersection points are given as follows:

$$\left.\overline{^{W}P_1{}^{W}P_{C_1}}\right|_{z=z_1^*} = \left.\overline{^{W}N}\right|_{z=z_1^*}, \qquad \left.\overline{^{W}P_2{}^{W}P_{C_2}}\right|_{z=z_2^*} = \left.\overline{^{W}N}\right|_{z=z_2^*} \quad (7)$$

where, z_i^* (i=1,2) denote the z-axis positions at which $\overline{^{W}P_i{}^{W}P_{C_i}}$ in (5) and $\overline{^{W}N}$ in (6) intersects. Eq. (7) provides four independent constraints totally because that the constraints of x and y axis in (7) are independent and *i=1, 2*. There are also four variables to be determined: z_i^* *(i=1,2)*, α, and β; thus, one can determine the variables using constraints in (7) (The detailed derivation is omitted due to the page limit).

After obtaining z_i^*, one can determine the intersecting positions as follows:

$$^{W}P_{Ti} = \left.\overline{^{W}P_i{}^{W}P_{C_i}}\right|_{z=z_i^*} \quad (8)$$

Finally, the 3D position of the target can be obtained by averaging the points in (8), as follows:

$$^{W}P = \left({}^{W}P_{T1} + {}^{W}P_{T2}\right)/2 \quad (9)$$

III. Kinematics using Weighted Damped Least Squares

When the target positions are given by the estimation algorithm, it is required that the kinematics scheme for the manipulator can provide proper solutions even when the motion profiles are out of the kinematic limitations. As mentioned in the introduction, the limitations regarding practical aspects are considered in this paper, as follows:

- the workspace limitation and singularity,
- maintenance of proper robot configuration after passing singular points, and,
- joint position limits.

A. Review on Weighted Damped Least Squares

By introducing the Jacobian matrix, one can describe the relationship between Cartesian space position, $\mathbf{x} \in \Re^n$, and joint space position, $\boldsymbol{\theta} \in \Re^m$, in the velocity level as follows [3]:

$$\dot{\mathbf{x}} = \mathbf{J}\dot{\boldsymbol{\theta}} \quad (10)$$

where, $\dot{\mathbf{x}} \in \Re^n$ denotes the velocity vector of the end-effector in Cartesian space; $\dot{\boldsymbol{\theta}} \in \Re^m$ the velocity vector in joint space; and $\mathbf{J} \in \Re^{n\times m}$ the Jacobian matrix. Without loss of generality, it can be assumed that $n = m = 6$ because the underwater manipulators commonly have six DOF to perform intervention tasks requiring spatial motion with translation and orientation.

By introducing the positive definite matrix, $\mathbf{W} \in \Re^{m\times m}$, one can describe (10) into weighted form as follows [18]:

$$\dot{\mathbf{x}} = \mathbf{J}_W\dot{\boldsymbol{\theta}}_W \text{, where,} \quad (11)$$

$$\mathbf{J}_W \equiv \mathbf{J}\mathbf{W}^{-1/2} \text{, and } \dot{\boldsymbol{\theta}}_W \equiv \mathbf{W}^{1/2}\dot{\boldsymbol{\theta}}. \quad (12)$$

By following the DLS formulation explained in [10], one can determine the cost function for the optimal kinematics problem as follows:

$$\min\left\| \begin{pmatrix} \mathbf{J}_W \\ \lambda\mathbf{I} \end{pmatrix}\dot{\boldsymbol{\theta}}_W - \begin{pmatrix} \dot{\mathbf{x}} \\ \dot{\mathbf{q}}_0 \end{pmatrix} \right\|^2, \quad (13)$$

where, $\dot{\mathbf{q}}_0$ denotes the negative gradient vector of an additional criterion, $h(\boldsymbol{\theta})$, to be minimized: i.e.,

$$\dot{\mathbf{q}}_0 = -k\nabla h(\boldsymbol{\theta}) = -k\frac{\partial h(\boldsymbol{\theta})}{\partial\boldsymbol{\theta}}. \quad (14)$$

The optimal solution minimizing square errors in (13) satisfies the following equation [12]:

$$\begin{pmatrix} \mathbf{J}_W \\ \lambda\mathbf{I} \end{pmatrix}^T \begin{pmatrix} \mathbf{J}_W \\ \lambda\mathbf{I} \end{pmatrix}\dot{\boldsymbol{\theta}}_W = \begin{pmatrix} \mathbf{J}_W \\ \lambda\mathbf{I} \end{pmatrix}^T \begin{pmatrix} \dot{\mathbf{x}} \\ \dot{\mathbf{q}}_0 \end{pmatrix}, \quad (15)$$

resulting the DLS solution of the weighted cost function in (13) as follows [10]:

$$\dot{\boldsymbol{\theta}}_W = \left(\mathbf{J}_W^T\mathbf{J}_W + \lambda^2\mathbf{I}\right)^{-1}\mathbf{J}_W^T\dot{\mathbf{x}} + \left(\mathbf{J}_W^T\mathbf{J}_W + \lambda^2\mathbf{I}\right)^{-1}\lambda\dot{\mathbf{q}}_0. \quad (16)$$

Let us define the damped pseudo inverse of $\mathbf{J}_W$ as follows:

$$\mathbf{J}_W^+ \equiv \mathbf{J}_W^T\left(\mathbf{J}_W\mathbf{J}_W^T + \lambda^2\mathbf{I}\right)^{-1}. \quad (17)$$

Then, the DLS solution in (16) can be described in the equivalent form as follows:

$$\dot{\boldsymbol{\theta}}_W = \mathbf{J}_W^+\dot{\mathbf{x}} + \left(\mathbf{I} - \mathbf{J}_W^+\mathbf{J}_W\right)\dot{\mathbf{q}}_0, \quad (18)$$

Because that, after some mathematical formulations, one can prove easily the followings:

$$\mathbf{J}_W^+ = \mathbf{J}_W^T\left(\mathbf{J}_W\mathbf{J}_W^T + \lambda^2\mathbf{I}\right)^{-1} = \left(\mathbf{J}_W^T\mathbf{J} + \lambda^2\mathbf{I}\right)^{-1}\mathbf{J}_W^T \text{, and} \quad (19)$$

$$\left(\mathbf{J}_W^T\mathbf{J}+\lambda^2\mathbf{I}\right)\lambda=\left(\mathbf{I}-\mathbf{J}_W^+\mathbf{J}_W\right). \quad (20)$$

Note that the solution in (18) provides more easy way to formulate the kinematics than (16), because, if the Jacobian, $\mathbf{J}$, is not singular, one can obtain the solution even when the manipulator has kinematic redundancy and the damping factor is set as zero: i.e., $n<m$ and $\lambda=0$; whereas, the solution in (16) does not give any proper solution because $\mathbf{J}_W^T\mathbf{J}_W$ has rank deficiency in the case. From (18) and (12), as a result, one can obtain the WDLS solution as follows:

$$\dot{\boldsymbol{\theta}}=\mathbf{W}^{-1/2}\dot{\boldsymbol{\theta}}_W \quad (21)$$

Note that the WDLS solution in (21) and (18) has three free-parameters that user can determine: i.e., λ, $h(\boldsymbol{\theta})$, and $\mathbf{W}$. Designing the free-parameters, one can obtain the kinematics solution optimizing (13).

From (17), one can find clearly that the roll of the damping parameter, λ, is to resolve the singularity problem. The inverse of $\mathbf{J}_W\mathbf{J}_W^T$ is not defined when $\mathbf{J}$ is singular. By adding non-zero damping parameter, in the case, one can obtain the weighted pseudo inverse in (17) properly.

By designing $h(\boldsymbol{\theta})$, one can optimize additional criteria using the null space of $\mathbf{J}_W$ as expressed in the second terms in the RHS of (18). Note that, when $n=m$ and $\lambda=0$, the $h(\boldsymbol{\theta})$ cannot influence to the kinematic solution because $\mathbf{J}_W^+\mathbf{J}_W=\mathbf{I}$ and the null space does not exist. If $\lambda\neq 0$, in contrast, there exist the null space because $\mathbf{J}_W^+$ in (18) cannot be the exact pseudo inverse of $\mathbf{J}_W$; thus, one can utilize $h(\boldsymbol{\theta})$ to design the kinematic solution even when $n=m$.

By designing weighting parameter, $\mathbf{W}$, one can obtain weighted square norm solution as used in energy minimization [16], torque minimization [17], and joint position limit avoidance [18]. As verified in [18], moreover, the utilization of $\mathbf{W}$ is more effective to avoid the joint limitation than that of kinematical null space design.

B. Optimal kinematics design for underwater manipulators

Utilizing the three free-parameters mentioned in the previous sub-section, one can design the kinematics of the manipulators based on the WDLS solution in (21) and (18).

Regarding the workspace limitation and singularity: Utilizing the damping parameter, λ, in (17), one can overcome the singularity problem; however, using constant λ may distort the kinematics solution because the parameter prevents $\mathbf{J}_W^+$ to be exact pseudo inverse of $\mathbf{J}_W$. For the exactitude of the inverse kinematics, therefore, λ has to be utilized only near the singularity as follows [15]:

$$\lambda^2=\begin{cases}\left(1-\dfrac{\sigma_{\min}}{\eta}\right)\lambda_{\max}^2 & ,\ \sigma_{\min}<\eta \\ 0 & ,\ \sigma_{\min}\geq\eta\end{cases} \quad (22)$$

where, $\sigma_{\min}$ denotes the minimum singular value of $\mathbf{J}_W$; $\lambda_{\max}$ the maximum value of λ; and, η the boundary of the singular region. By designing $\lambda_{\max}$ and η appropriately, one can resolve the singularity problem, maintaining the exactitude of the inverse kinematics when $\mathbf{J}_W$ is not singular.

Note that the WDLS in (21) with the damping parameter in (22) can resolve the workspace limitation problem. The cases when the manipulators are fully extracted or retracted are also ones of the singular positions and the WDLS can resolve the kinematics even when the cases occur. When the target position is out of workspace, moreover, the WDLS can provide proper kinematics in the view point of the weighted least squares that minimize the cost function in (13): i.e., the WDLS provides the nearest position to the target, even though the target positions are not reachable.

Regarding maintenance of proper robot configuration after passing singular points: The robot configuration can be changed to unwanted posture after passing the singular points. The robot posture deciding an end-effector position is not unique as explained in Chapter 4 in [3]; when designing the motion of the manipulator, user chooses preferred one among the postures. When kinematics is resolved with numerical approaches, however, the robot configuration can be changed after passing singularity because, at the singular point, it is possible to move to any posture and the numerical method does not know which posture is preferred.

The free-parameter, $h(\boldsymbol{\theta})$, can be utilized to indicate the preferred posture when the manipulator is in singular position: i.e., by designing $h(\boldsymbol{\theta})$ to minimize the error between the preferred posture and current one, one can make the manipulator maintain the preferred posture. For example, one can set the preferred posture as the one which is far from joint position limit: in the case, $h(\boldsymbol{\theta})$ can be designed as follows:

$$h(\boldsymbol{\theta})=\sum_{i=1}^{m}\left(\frac{\theta_i-\bar{\theta}_i}{\theta_{i,\max}-\theta_{i,\min}}\right)^2, \quad (23)$$

where, $i=1,\ldots,m$; θ_i denotes elements of $\boldsymbol{\theta}$; $\theta_{i,\max}$ and $\theta_{i,\min}$ the maximum value and the minimum value of θ_i, respectively; and, $\bar{\theta}_i$ signifies the medium between $\theta_{i,\max}$ and $\theta_{i,\min}$ as follows:

$$\bar{\theta}_i=\frac{\theta_{i,\max}+\theta_{i,\min}}{2}. \quad (24)$$

The shape of (23) is illustrated in Fig. 2(a), which shows that $h(\boldsymbol{\theta})$ has the minimum value when $\theta_i=\bar{\theta}_i$, and monotonically increase when θ_i approaches $\theta_{i,\max}$ or $\theta_{i,\min}$.

As explained in (14), the negative gradient of (23) is obtained as follows:

$$\dot{\mathbf{q}}_0=-k\left[2\frac{q_1-\bar{q}_1}{\left(q_{1,\max}-q_{1,\min}\right)^2},\ \ldots,\ 2\frac{q_m-\bar{q}_m}{\left(q_{m,\max}-q_{m,\min}\right)^2}\right]^T. \quad (25)$$

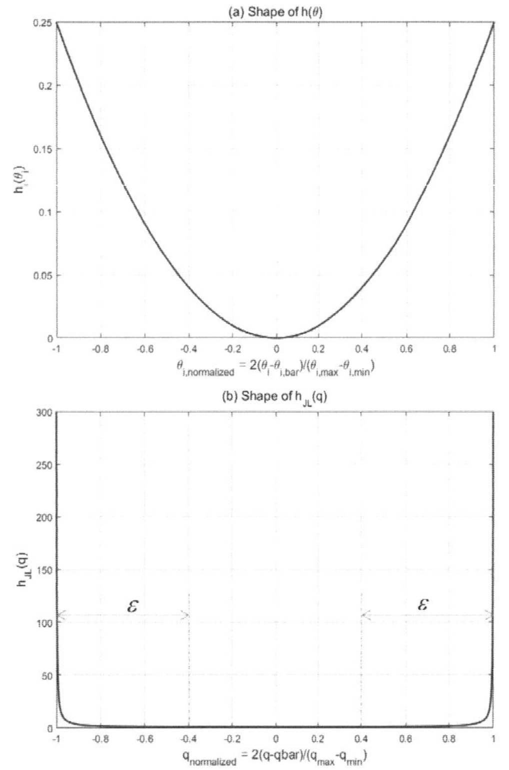

Fig. 2 Shape of cost functions in (24) and (27)

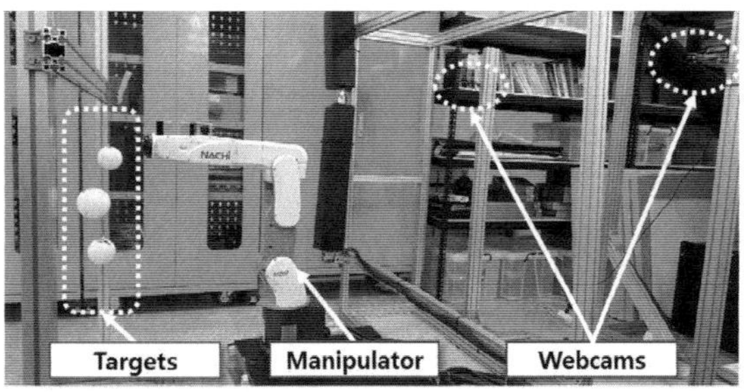

Fig. 3 Experimental Setup

Regarding joint position limits: As explained in [18], one can utilize the weight matrix, $\mathbf{W}$, to avoid to meet the joint limits. In this paper, the position limit each joint are considered. Here, a cost function is introduced to design the elements of $\mathbf{W}$ as follows:

$$h_{JL}(q)=\begin{cases}\dfrac{\varepsilon^2}{(q_{\min}+2\varepsilon-q)(q-q_{\min})} & ,\ q_{\min}\le q<q_{\min}+\varepsilon\\ 1 & ,\ q_{\min}+\varepsilon\le q\le q_{\max}-\varepsilon\\ \dfrac{\varepsilon^2}{(q_{\max}-q)(q-q_{\max}+2\varepsilon)} & ,\ q_{\max}-\varepsilon<q\le q_{\max}\end{cases}. \quad (26)$$

where, q denotes an input variable; $q_{\min}$ and $q_{\max}$ the minimum value and the maximum value of q, respectively; and, ε signifies the boundary for the region near the limits. The shape of (26) is depicted in Fig. 2(b), which shows that the function in (26) remains unity within the region far from the limits and increase rapidly when q approaches the limits. The gradient of (26) is obtained as follows:

$$\frac{\partial h_{JL}(q)}{\partial q}=\begin{cases}\dfrac{2\varepsilon^2(q-q_{\min}-\varepsilon)}{(q_{\min}+2\varepsilon-q)^2(q-q_{\min})^2} & ,\ q_{\min}\le q<q_{\min}+\varepsilon\\ 0 & ,\ q_{\min}+\varepsilon\le q\le q_{\max}-\varepsilon\\ \dfrac{2\varepsilon^2(q-q_{\max}-\varepsilon)}{(q_{\max}-q)^2(q-q_{\max}+2\varepsilon)^2} & ,\ q_{\max}-\varepsilon<q\le q_{\max}\end{cases}. \quad (27)$$

Utilizing (27), one can design the weight matrices, $\mathbf{W}$, as following diagonal matrices:

$$\mathbf{W}=diag(w_i) \quad (28)$$

where, the diagonal elements of each matrix are designed as follows [18]:

$$w_i=1+\left|\frac{\partial h_{JL}(\theta_i)}{\partial \theta_i}\right|. \quad (29)$$

IV. Experimental Validation

The performance of the proposed assistant technique is verified through laboratory experiments. Fig. 3 shows the experimental setup which consist of an industrial manipulator, two webcams, and targets.

Experiments are repeated five times, and their results are arranged in Fig. 4 ~ Fig. 6 and Table 1: Fig. 4 and Table 1 show the results of 3D position estimation, and Fig. 5 and Fig. 6 illustrate the robot motion according to kinematics algorithm.

From the Fig. 4 and Table 1, one can find that the estimation algorithm has error of couple of centimeters which can be caused by calibration errors of cameras and deployments, or inaccuracy in touch screen input. The results show that the estimation algorithm has enough accuracy to approach 50~100mm boundary near the target.

From Fig. 5 and Fig. 6, one can note that the kinematics algorithm works correctly to approach the target accurately (in the case of Ball 1 and Ball2); moreover, the algorithm works stably even when the target position is out of workspace (in the case of Ball 3). Note that, in the case of Ball 3, the robot cannot reach the target position even though it stretches fully. Fig. 5 and Fig. 6 verify that the kinematics algorithm provides a solution that minimize the norm error in (13) which means the nearest position to the target.

V. Conclusion

In the paper, a vision based assistant technique for underwater manipulations is proposed. Especially, 3D target position estimation issue and inverse kinematics issue are addressed. Regarding target position estimation issue, a technique which can estimate 3D position of the objects utilizing 2D position information on the two images from cameras deployed in stereo is addressed. Via touch screen, the 2D position information is fed by operators for guaranteeing reliability of the estimation results. Regarding kinematics issues, a kinematics solution based on the weighted damped least squares is investigated to overcome the limitations on the manipulators. By designing appropriate cost functions, the proposed kinematic solution provides a solution optimized in the manner of weighted norm, even when the manipulators meet those limitations. Through the laboratory experiments, the performance of the proposed assistant technique is verified.

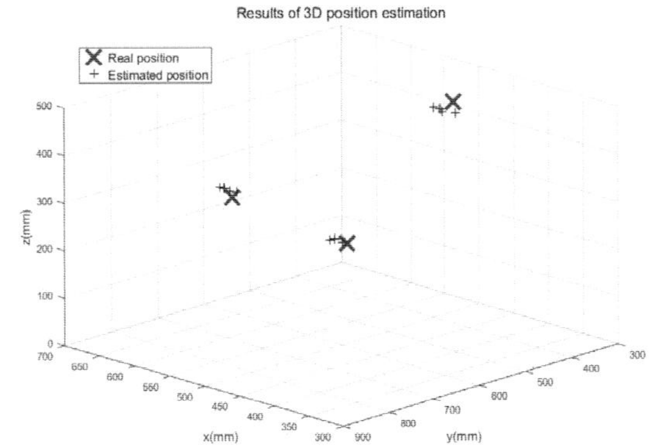

Fig. 4 Target position estimation: Results of 5 trials

TABLE I
RESULTS OF 3D POSITION ESTIMATION

		Real Position (mm)	Average of estimated position (mm)	Average of estimated error (mm)
	x	500.0	516.9	-16.9
Ball1	y	355.0	353.3	1.73
	z	440.0	415.2	24.8
	x	500.0	501.9	-1.90
Ball2	y	590.0	607.5	-17.5
	z	210.0	220.2	-10.2
	x	500.0	502.1	-2.08
Ball3	y	840.0	848.3	-8.33
	z	380.0	398.0	-18.0

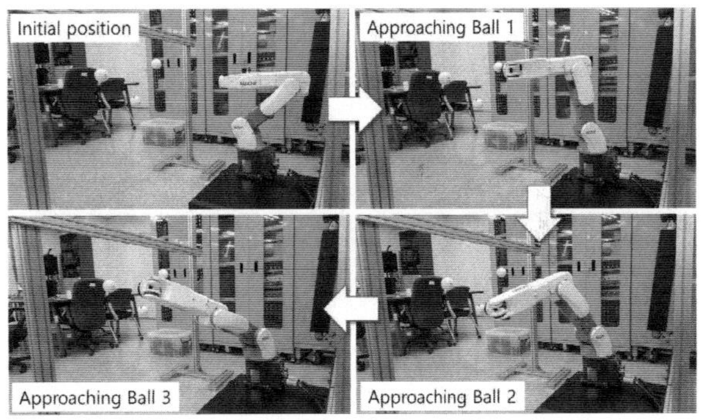

Fig. 5 Robot posture: Results of 1st trial

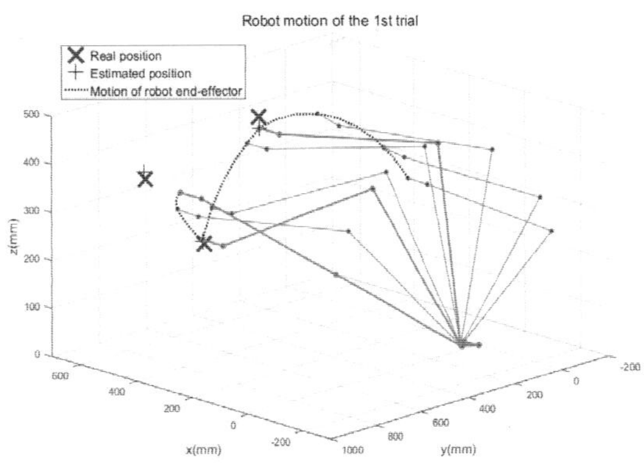

Fig. 6 Robot motion: Results of 1st trial

REFERENCES

[1] Y. J. Lee, J. W. Choi, H. T. Choi, 2014, "Underwater Robot Localization by Probability-based Object Recognition Framework Using Sonar Image", *Journal of Korea Robotics Society*, vol.9, no.4, pp.232-241.

[2] David A. Forsyth, Jean Ponce, 2003, *Computer Vision A Modern Approach*, Pearson Education International.

[3] John J. Craig, 1989, *Introduction to Robotics*, Addison-Wesley, 2nd ed..

[4] Richard Hartley, Andrew Zisseman, 2003, *Multiple View Geometry in computer vision*, CAMBRIDGE, 2nd ed..

[5] R. D. Christ, R. L. Wernli, 2014, The ROV manual: A user guide for remotely operated vehicles, Elsevier, 2nd ed.

[6] M. Hildebrandt, J. Albiez, F. Kirchner, 2008, "Computer-based control of deep-sea manipulators," Oceans 2008 - MTS/IEEE Kobe Techno-Ocean, pp. 1-6.

[7] G. Marani, S. K. Choi, J. Yuh, 2009, "Underwater autonomous manipulation for intervention missions AUVs," Ocean Engineering, vol. 36, iss. 1, pp. 15-23.

[8] M. Prats, J. C. Garcia, S. Wirth, D. Ribas, P. J. Sanz, P. Ridao, N. Gracias, G. Oliver, 2012, "Multipurpose autonomous underwater intervention: a systems integration perspective," 20th Mediterranean Conf. on Control & Automation, pp. 1379-1384.

[9] J. C. Garcia, J. J. Fernandez, P. J. Sanz, R. Martin, 2010, "Increasing autonomy within underwater intervention scenarios: the user interface approach," IEEE Int. Systems Conference, vol. 4, pp. 71-75.

[10] A. S. Deo, I. D. Walker, 1995, "Overview of damped least-squares methods for inverse kinematics of robot manipulators," J. of Intelligent and Robotic Systems, vol. 14, pp. 43-68.

[11] S. R. Buss, 2004, "Introduction to inverse kinematics with jacobian transpose, pseudoinverse and damped least squares methods," [online] Available: http://web.cse.ohio-state.edu/~parent/classes/694A/Lectures/Material/IKsurvey.pdf

[12] C. W. Wampler, 1986, "Manipulator inverse kinematic solutions based on vector formulations and damped least-squares methods," IEEE Trans. on Systems, Man, and Cybernetics, vol. 16, no. 1, pp. 93-101.

[13] Y. Nakamura, H. Hanafusa, 1986, "Inverse kinematics solutions with singularity robustness for robot manipulator control," J. of Dynamic Systems, Measurement, and Control, vol. 108, pp. 163-171.

[14] S. Chiaverini, B. Siciliano, O. Egeland, 1994, " Review of the damped least-squares inverse kinematics with experiments on an industrial robot manipulator," IEEE Trans. on Control Systems Technology, vol. 2, no. 2, pp. 123-134.

[15] S. Chiaverini, 1997, "Singularity-robust task-priority redundancy resolution for real-time kinematic control of robot manipulators," IEEE Trans. on Robotics and Automation, vol. 13, no. 3, pp. 398-410.

[16] D. E. Whitney, 1972, "The mathematics of coordinated control of prosthetic arms and manipulators," ASME J. of Dynamic Systems, Measurement, and Control, vol. 94, no. 4, pp. 303-309.

[17] J. M. Hollerbach, K. C. Suh, 1987, "Redundancy resolution of manipulators through torque optimization," IEEE J. of Robotics and Automation, vol. 3, no. 4, pp. 308-316.

[18] T. F. Chan, R. V. Dubey, 1995, "A weighted least-norm solution based scheme for avoiding joint limits for redundant joint manipulators," IEEE Trans. on Robotics and Automation, vo. 11, no. 2, pp. 286-292.

[19] J. Yuh, 2000, "Design and control of autonomous underwater robots: a survey," Autonomous Robots, vol. 8, pp. 7-24.

[20] M. R. Dhanak, N. I. Xiros (Eds.), 2016, Springer Handbook of Ocean Engineering, Springer.

A General Overview of Compressed Air Testing for Purposes of Improving Commercial Diving Safety

TRI Air Testing, Inc.

Austin, Texas

January 2017

Table of Contents

Disclaimer

This document has been reviewed in accordance with TRI Air Testing, Inc. policies and is approved for publication. Any mention of trade names or commercial products does not constitute endorsement or recommendation for use.

List of Tables

List of Figures

Preface

This report, *A General Overview of Compressed Air Testing for Purposes of Improving Commercial Diving Safety*, is the product of TRI Air Testing's forty -years of industry knowledge. Conforming to air quality specifications of many regulatory agencies and industries, along with developing and customizing our own analytical methods, TRI Air Testing, Inc. is the industry leader in this type of analysis.

This report will detail the components that are found in breathing air, the components and contaminates we test for, and why we test for these particular analytes as they relate to both the commercial and recreational dive industries. This report is designed to address the following components, contaminates, and physical properties, that all combined can make up the majority spectrum of analytes that need to be tested in order for breathing air to be considered safe for usage:

- **Oxygen (O_2)**
- **Nitrogen (N_2)**
- **Carbon Monoxide (CO)**
- **Carbon Dioxide (CO_2)**
- **Methane (CH_4)**
- **Total Hydrocarbons (THC) minus Methane (CH_4)**
- **Oil Mist and Particulates (condensed hydrocarbons)**
- **Mold and Bacteria**
- **Water Vapor (moisture content)**
- **Dew Point**
- **Pressure Dew Point**

All the above will be examined and discussed in detail in the following paragraphs of this report. The report will hopefully facilitate dialogue and show the need for continued and improved safety as it relates to the testing of breathing air being used.

Authors and Contributors

Jason Vandagriff, BS, *Associate Laboratory Director TRI Air Testing, Inc.*

Jonathon F. Sauseda, MBA, *Vice President TRI Air Testing, Inc.*

Acknowledgements

The author would like to acknowledge ***Dr. Edward Golla, PhD*** for his contributions to TRI Air Testing, Inc. since its inception over forty years ago. All of TRI's analytical methods were developed by Dr. Golla. His expertise in the usage of and testing of compressed gases is unrivaled.

Introduction

For the purposes of this report, all analytical discussion will be focused on what is considered a breathing gas. A breathing gas is any gas used for respiration purposes. Air is the most common and natural type of breathing gas. Air is made up of approximately 21% oxygen (O_2), 78% nitrogen (N_2), and 1% of various trace elements, with the most abundant of those trace elements being argon (Ar). Breathing gases are pure gases or blended gases that are used to support life in certain enclosures and breathing equipments such as SCBA, SCUBA, surface supplied diving, submarines, space vessels and habitats, medical life support and anesthetic usage, and mountaineering. Since we are discussing breathing gas as it is related to diving usage, we will be discussing the functions of the gas in a hyperbaric state, or in a state of pressure being greater than sea level atmospheric pressure (1atm). Hyperbaric breathing gas can be supplied for diving from the surface (umbilical) and from SCUBA (self contained underwater breathing apparatus). A hyperbaric breathing gas must contain a few basic features: [1][2][3]

- **The breathing gas must contain enough oxygen to support the working rate of the breather.**
- **Contaminates, whether they are solid particulates or other potential harmful gases (carbon monoxide and carbon dioxide) must be at very small concentrations to be considered safe for usage.**
- **The breathing gas must not become toxic when underwater in a higher pressure state. Both oxygen and nitrogen can become toxic at certain depths and pressures.**

The diving industry uses regular natural air or synthetic air as a starting point to blend gases according to the pressure or depth at which the diver or user will be operating. As stated, oxygen and nitrogen can become toxic at certain pressures. Narcosis, also known as nitrogen narcosis, is an anesthetic effect of certain gases at pressure. Narcosis can produce a state similar to the effects of alcohol intoxication. Depending on the diver, narcosis can occur in shallow dives but is rarely seen in dives of less than forty meters (130 ft). [1] [3] With the exception of helium and neon, all gases that can be breathed have the potential to cause narcosis. The degree at which the gas can cause narcosis is based on it's lipid solubility, which is basically a factor of how well the gas can absorb into a diver's tissue. [3] Another more dangerous health issue involving diving is oxygen toxicity. The result of breathing oxygen at higher partial pressures is called hyperoxia, or an excess of oxygen in the body tissue. In this case, too much of a good thing is a bad thing. Oxygen toxicity can affect all cells of the human body. The cells in central nervous system, pulmonary (lung), and ocular (eye) are the three areas of most concern when dealing with this. [7] The biggest fear of oxygen toxicity is a potential seizer that can result from the effects. Selecting the correct concentrations of oxygen enriched or oxygen reduced gas blends is one of the most important and calculated decisions facing any diver that is not diving at recreational depths. The most common blends to combat the narcotic effects of gases at pressure are Nitrox, Trimix, Heliox, Hydrox, Hydreliox, and Heliair.

Each one of these breathing gas blends are useful at varying depths (pressures), factoring in the length of the duration of the dive as well. For the purposes of this report, these blends will not be discussed in detail. It is important to know however, that for any breathing gas that it being used, whether it is air or one of the above mentioned gas blends, the quality of the gas must be ensured for the safety of the users. This is where a testing facility such as TRI Air Testing, Inc. comes in to play. In the following paragraphs, this report will detail what is tested and how it is tested in regards to compressed gas analysis. TRI, along with other similar air testing laboratories, will send out air testing kits with sampling media for onsite sampling performed by the customer. The customer will then send back the kit (if the customer didn't purchase it for ownership) and the sampling media to the air testing laboratory for analysis. The air testing kits contain an air sampler. This air sampler, along with pressure fittings, attaches directly to the air compressor or gas source being tested. The sampling media is contained within or attached to the air sampler. An example of an air sampler (TRI Air Testing , Inc. refers to it as a flow section) is seen below in **Figure 1**.

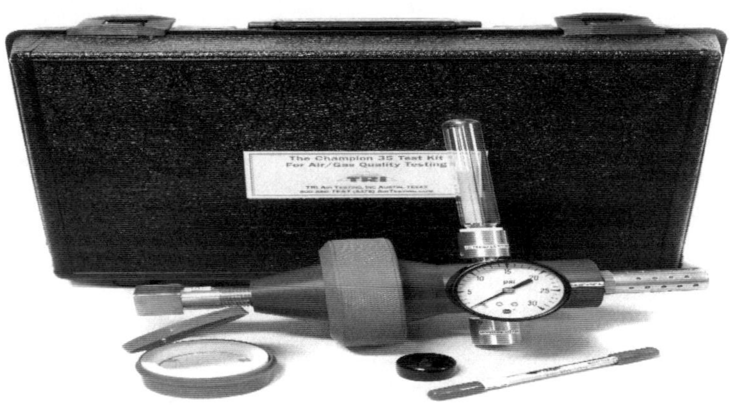

Figure 1. A TRI Air Testing, Inc. Champion Flow Section kit with oil mist and particulate filter, a Drager tube for moisture testing, and a 60 ml glass bottle for gas sampling.

The filter used for oil mist and particulate analysis, the Drager tube used for moisture testing and nitrous fume testing, and the filter used for mold and bacteria analysis are all sampled and analyzed based on a function of air volume. TRI Air Testing Inc. will base all data reporting limits and data calculations from pressure and time. By capturing and reporting the pressure (caused by varying sized orifice plates contained in the air ampler) and air flow time, an accurate volume of air can be calculated. Oil mist and particulate data reporting limits, moisture, and mold and bacteria can then be calculated based off the volume of air that passes through these sampling media. The gas analysis of percent volume of oxygen and nitrogen along with the parts per million (ppm) concentrations of total hydrocarbons, carbon monoxide, and carbon dioxide are all obtained by pressurizing the gas sample bottle or cylinder that is sent out with the test kit. The actual methods to which all of these analytes are analyzed at the air testing laboratory will be detailed in the following pages of this report.

Analysis of Gaseous Contaminants

The gases would be analyzed by **gas chromatograph** (GC) using both **thermal conductivity (TCD)** and **flame ionization (FID)** detectors (**Table 1**). At TRI, this dedicated GC system is used exclusively for analysis of compressed air samples. Using appropriate plumbing within the GC, a single injection directly from the sample provides a complete gas analysis. The sample information and analytical data is accumulated into a detailed printout of results that is generated into a report that customers can access online.

(TABLE 1) **TRI AIR TESTING, INC. ANALYTICAL METHODS**

Analyte	*Analysis Method*	*TRI's Level/Accuracy*	*TRI's Detection/Reporting Limits*	*TRI's Recommended Maximum Allowable Level for Compressed Breathing Air*
Oxygen and Nitrogen	Gas chromatography with thermal conductivity detector	+\- 0.4% absolute or +/- 2% of concentration	0.5%	19-22% O_2 and 77-80% N_2 (for non blended gases)
Carbon monoxide	Gas chromatography with catalytic converter followed by flame ionization detector	+\- 5% of reported value or 1 ppm whichever is greater	0.5 ppmv	10 ppmv
Carbon dioxide	Gas chromatography with catalytic converter followed by flame ionization detector	+\- 5% of reported value or 20 ppm whichever is greater	20 ppmv	1000 ppmv
Total gaseous hydrocarbons (methane)	Gas chromatography with direct flame ionization detector	+/- 5% of reported value or 1 ppm whichever is greater	1 ppmv	5 ppmv
Condensed Hydrocarbons (Oil Mist and Particulate)	Standard gravimetric, with hexane extraction for oil mist if results are with 90% of the specification	+/- 0.1 mg/m^3 or 10%, whichever is greater	0.1 mg/m^3 (can be as low as 0.01 mg/m^3 based on volume)	0.1 mg/m^3
Moisture Test	Color indicator tube with critical orifice size to measure air flow (test performed on site)	+/-30% or 1 ppm whichever is greater	2 ppmv	25 ppmv

There are many sound analytical techniques used for testing moisture. However, an air sample from an air compressor or gas source cannot be transported back to an air testing laboratory with the same moisture content as when it was sampled without used a very heavy and expensive high-pressure cylinder.

Therefore, TRI uses a color indicator Drager tube to allow the moisture content in the source sample to be accurately measured at the time of sampling. Drager tubes are also used to detect for the presence of sulfur fumes and nitrous fumes as well, however those fume analysis are not typical in the diving industry and are not part of the standard catalog of breathing gas analysis.

TRI has selected GC as the most appropriate method for the analysis of gaseous breathing air components. In GC, the components of the air sample are carried through an analytical column by an inert gas (the carrier gas). The sample mixture is partitioned between the moving carrier gas and a solid support and/or stationary phase. The solid support stationary phase selectively retards the sample components according to their distribution coefficient until they form separate bands in the carrier gas. These component bands leave the column in the gas stream and are sensed by the appropriate detector (FID or TCD). The resulting analog signal is input into an integrator, recorded on an electronic data file, and printed by the attached printer, thus creating a physical copy as well as an electronic file. The integrator records the retention times of the peaks (which identify the components) and area under the peak (which corresponds to the amount present).

TRI's GC is an SRI Model 8610A and has been modified so that all gas analyses are performed with one injection using the sample itself as the injector instead of a syringe. Sample loops are used with a multi-position valve to withdraw a sample from the sample and introduce it into the GC. Techniques for the analysis of different air components or contaminants are described in **Table 2**.

(TABLE 2) **TRI Techniques for the Analysis of the Different Air Components or Contaminants**

Air Component/ Contaminants	Techniques for Analysis
Oxygen and Nitrogen	Oxygen and Nitrogen pass through a mol sieve column and are detected by a thermal conductivity detector (TCD).
Carbon monoxide, Carbon dioxide and Methane	Carbon monoxide, carbon dioxide, and methane from the sample aliquot pass through an analytical column that separates the three analytes in time. A methanizer then converts carbon monoxide and carbon dioxide to methane (and leaves the methane as methane). Once reduced, the three components pass into a flame ionization detector (FID) that produces peaks on a chromatogram, which identifies the components by retention time and concentration measured.
Total Gaseous Hydrocarbons	All gaseous hydrocarbons are allowed to elute very quickly and will pass to the FID at one time and be recorded as one entity. Carbon monoxide, carbon dioxide, oxygen, and nitrogen do not provide any response on an FID, so it is not necessary to correct resultant values for their presence.

Oxygen

Oxygen is the primary element that supports life and cellular respiration. The diatomic form of oxygen (O_2) makes up about 20.8% of the Earth's atmosphere and in other oxide forms, the element makes up about half of the Earth's crust [4]. As stated earlier, the concentration of O_2 in the breathing gas is one of the most important factors in the diving industry from a safety standpoint. The Occupational Safety and Health Administration (OSHA) has determined that the optimal breathing range to be between 19.5 and 23.5 percent O_2 under normal atmospheric pressure. It has been determined that the minimum O_2 concentration in any pure form or mixture needs to be at least 14% to sustain a typical human body. In percent volume concentrations of oxygen below 10%, the human body will typically become severely compromised [3].

TRI Air testing, Inc. test to specifications for numerous industries and each industry is governed by different regulatory agencies. TRI Air Testing, Inc uses CGA (Compressed Gas Association) G-7.1 Grade E 2011 edition, the Navsea 09110-LP-106-0957 USN Diving Manual, Vol1, Rev 6 15Apr2008, TBL 4-1, and the Association of Diving Contractors International Consensus Standards for Commercial Diving and Underwater Operations 6.1 Edition 2014, as the primary specification limits for the diving industry air quality analysis. TRI also test to the British and Canadian diving specifications as well. The specification limits for percent volume of oxygen from a gas source for these regulatory entities lies between 19.5-23.5%, with most being 20-22%. TRI Air Testing, Inc. can, however, customize specifications to meet our customer demands and allow for an oxygen "pass spec" grade should the gas require that the oxygen percent volume be outside these ranges. For example, TRI Air Testing, Inc. has developed a Nitrox gas specification that follows the standard Grade E spec limits for all other analytes except oxygen. In this case, the Nitrox specification allows for a 24-39% oxygen range, which gives our customers a larger window to accommodate the varying Nitrox membrane blends. A list of common specification limits for breathing air can be found on the TRI Air Testing, Inc. website at www.airtesting.com.

Nitrogen

The diatomic form of nitrogen (N_2) forms about 78% of the Earth's atmosphere, making it the most uncombined element on the planet. [3] The Human body contains a little over 3% nitrogen, making it the forth most abundant element of the human body. [3] Nitrogen is a critical component in cellular energy transfer in all biological organisms. The nitrogen cycle is a homeostasis process that details the flow of the element from the air into the biosphere and organic compounds, then back into the atmosphere. Being that it constitutes the majority volume of standard atmospheric air, it is a critical analyte for air quality testing. The regulatory agencies that published the specifications for the quality breathing air standards do not have upper and lower specification limits for the percent volume of N_2. It is assumed that in standard atmospheric compressed breathing air, the N_2 will be in balance with the percent volume of O_2. If the critical respiration element of oxygen (O_2) volume is within the specification limits, it is assumed N_2 will be in the proper volume balance of approximately 78%. As stated above, the diving gas blends will increase or decrease the O_2 and N_2 as needed to combat the narcotic effects of these gases. In some cases where the pressures will be very high, N_2 will be removed completely from the diving gas blend.

Total Hydrocarbons Including Methane

A hydrocarbon in organic chemistry is any compound containing only hydrogen and carbon. The majority of naturally occurring hydrocarbons are contained in crude oil, where decomposed organic material provides an abundance of hydrogen and carbon atoms. This abundance of carbon and hydrogen atoms can then bond together and create a limitless source of bonded chains. [6] Hydrocarbons can be found in all the main physical states of solid, liquid, and gaseous forms. The combustion of hydrocarbons is the primary source of energy on Earth. The combustion of hydrocarbons and their virtual unlimited usage in a variety of different industrial processes, make them a key contaminate in breathing gases. The hydrocarbon content must be accounted for when performing any compressed air quality analysis. Hydrocarbon toxicity to the human body can be extremely varied depending on the species of the hydrocarbon.[6] Typically a high exposure of hydrocarbon toxicity is a result of inhalation. This can cause varied respiratory system and central nervous system problems in the human body. Excess "oil mist" (some semi-volatile and condensed hydrocarbons) deposits in the alveoli can cause an intense inflammation know as lipoid pneumonia. This can cause alveoli to dilate and rupture, decreasing the total surface area available for the transfer of oxygen and carbon dioxide known as emphysema.

At TRI, the volatile and some semi-volatile hydrocarbons that are not condensable, minus methane (CH_4), are allowed to elute very quickly in our GC-FID instrument and are combined into one fully quantifiable chromatography peak. This is referred to as the total hydrocarbon (THC) peak. The methane analyte elutes shortly after the THC peak and is quantified separately for the purposes of data reporting. The condensable hydrocarbons and some semi-volatile hydrocarbons are analyzed separately. These compounds are referred to as "Oil Mist" and they are tested gravimetrically at TRI Air Testing, Inc. Both of these GC and gravimetric analysis will ensure that there is or is not a presence of an organic solvent in a compressed gas system. For the purposes of breathing air specifications, the regulatory agencies that TRI test to put an upper specification limit of 25 ppmv for total volatile hydrocarbons including and excluding methane. The upper specification limit for oil mist by the regulatory diving agencies and OSHA is 5 mg/m^3. OSHA has an extremely varied eight-hour daily exposure limit of volatile hydrocarbon species based on their individual toxicity. TRI has a minimal reporting limit of 1 ppmv for THC including methane and the majority of TRI's compressed air analysis report THC and methane results of less than 2 ppmv. If a customer's gas results for THC and methane are shown to be elevated (above 15 ppmv), a TRI senior scientist will report an immediate concern to the customer and suggest to the customer that their air source be retested and possibly evaluated on a GC-MS for hydrocarbon speciation.

Carbon Monoxide

Carbon monoxide (CO) is the deadliest of the toxic gases commonly found in compressed air. The fact that it is colorless and odorless makes it virtually impossible to detect when breathing. CO combines readily with red blood cells and prevents the transfer of oxygen to the tissues causing oxygen starvation (hypoxia). The physical effects of CO poisoning are severe headaches, heart arrhythmia, confusion, loss of equilibrium, unconsciousness, and eventually death. Carbon monoxide forms when there is a situation where carbon compounds are being oxidized and there is not enough oxygen to form carbon dioxide.[5] This can commonly occur in enclosed spaces where there is a combustion of carbon containing compounds. Worldwide, the largest source of carbon monoxide is natural in origin due to chemical reactions in the lower portion of the Earth's atmosphere (troposphere) that generates about 5×10^{12} kilograms per year.[4] According to the Compressed Gas Association (CGA) and OSHA, carbon monoxide in compressed air should be limited to 10 ppm. This limit is set for CGA Grade D breathing air for industry and general respiratory use. For sport scuba diving up to 125 feet as well as commercial diving, the limit is also at 10 ppm (CGA Grade E). TRI will consider a CO value above 10 ppm an immediate concern and the customer will be notified immediately. The majority of compressed gas samples that TRI analyze are below the reporting limit of 0.5 ppmv for carbon monoxide.

Carbon Dioxide

Carbon dioxide (CO_2) is also classified as a toxic gas, but can be tolerated by the human body at much higher concentrations than carbon monoxide. Even though it is considered toxic, it is an essential life sustaining component. [7] Carbon dioxide is present naturally in the Earth's atmosphere at about 450 ppmv and is the main carbon source for life on Earth. [4] Carbon dioxide is an essential component in photosynthesis, which is the conversion of light energy into chemical energy by plants and some organisms. The byproduct of photosynthesis is oxygen. The CGA Grade D and Grade E upper specification limit is 1000 ppmv, where OSHA allows a daily exposure limit of up to 5000 ppmv. TRI recommends a 1000 ppmv CO_2 upper limit for breathing gas. The effects of carbon dioxide poisoning are very similar to those of carbon monoxide; however, first it stimulates the respiratory center which causes an increased rate and depth of breathing. This is especially dangerous, not only because it speeds up the effects of CO_2, but also increases the intake of other contaminants that may be present.

Both carbon monoxide and carbon dioxide can't be detected directly with a flame ion detector (FID). An FID detector is limited in detecting oxygenated species. In order for these two compounds to be analyzed by a GC-FID, the two compounds must first go through a catalytic conversion to methane. TRI uses a bed of nickel-62 (^{62}Ni) as the methanizer for these two compounds. These two reduced compounds including methane itself are separated by retention times and can be quantified separately.

Moisture and Dew Point

Moisture (water vapor or condensed water) in compressed air is as much a contaminant as others but for a different reason. In cold climates, excess water vapor can cause the regulator or valve to freeze up and cause the SCBA (self-contained breathing apparatus) to become inoperative, thus negating its life-supporting purpose. Excess moisture can promote cylinder corrosion, rust, and may diminish the effectiveness of the catalyst in a purification system which converts CO to CO_2. Moisture in the breathing gas itself is not dangerous to the human body, just the equipment. Regular sea level atmospheric air contains anywhere from 13,000 ppmv to 20,000 ppmv depending on the temperature and absolute humidity at any given time or location. The CGA states that "water content of compressed air required for any particular grade may vary with the intended use from saturated to very dry. For breathing air used in conjunction with self-contained breathing apparatus in extreme cold where moisture can condense and freeze causing the breathing apparatus to malfunction, a dew point not to exceed -65°F (24 ppmv) or 10°F lower than the coldest temperature expected in the area is required."[1] Also compressed air which is to be used at temperatures below freezing should have excess water vapor removed so as to attain a dew point below the minimum temperature anticipated.

The above statement by the CGA addresses the term "dew point". The dew point is the temperature at which the air can no longer hold the amount of water vapor present and the water vapor will condense into droplet form. These droplets can form fog, clouds, or actual droplets of water on any ground level surface that has surface tension. The dew point is a function of temperature and absolute humidity. Warmer air is less dense and has the capacity to hold more water vapor than denser colder air, so it takes a much lower absolute humidity and a smaller temperature differential for the water vapor to condense out of colder air. This is why the upper specification Navy limit is very strict when it comes to moisture and dew point. Their specification limits require very dry air (less than 24 ppmv of water vapor) to accommodate very cold regions that compressed breathing gas can be used.

TRI reports to customers the dew point data at standard atmospheric pressure. There is a difference between this dew point and the actual dew point that is in a compressed gas system. In a compressed gas system the "pressure dew point" must be accounted for. When any gas, including air, is compressed it becomes heated. As stated above, warmer air can hold much more water vapor. As a result, compressed air always has a relative humidity of 100% and a resulting higher pressure dew point than the corresponding atmospheric dew point. For example, a compressed air system that is being analyzed has a moisture content of **648 ppmv** with a corresponding atmospheric dew point of **-13°F**. The compressor itself, however, is being operated at **1115 psig**, which would give a pressure dew point of **30°F** at that same **648 ppmv** water vapor content**.** The regulatory agencies set their specifications at a standard atmospheric dew point, but it is important to always keep in mind the corresponding pressure dew point that is in play with any compressed gas/air system. Compressed air systems used for breathing gas have mechanical or chemical dryers in place. If working properly, these dryers will keep the moisture content to extremely low amounts. A properly functioning desiccant dryer, coupled with a properly working coalescing filter, can produce an atmospheric dew point as low as -95°F.

Oil Mist and Particulates

Condensed hydrocarbons were previously discussed in this report as to why they are part of the standard breathing gas analysis from a health standpoint. The regulated upper limit for breathing gas condensed hydrocarbons is 5 mg/m^3. Most properly working compressed gas sources will generate no detectable amount of oil mist.

The same 5 mg/m^3 upper spec limit is also in place for solid particulate matter as well. At TRI, a combined amount of oil mist and particulate is reported to the customers, as well as the amounts of each component separated out individually. The oil mist weight is accurately recorded by extracting out the condensed hydrocarbons with a sample filter wash of hexane. The weight that is lost on the sample filter after the hexane extraction is considered the oil mist weight or the oil mist amount that is present. Compressed gas/air systems will have particulate filtration mechanisms in place to filter out particles of anything above very small micron sizes. Particulates at micron and sub micron sizes don't necessarily pose a health threat to the human body. Humans breathe in billions of micron sized dust and pollen particles everyday in the course of normal respiration. Particulate matter does pose a similar problem as moisture however, as they can build up over time and diminish the functionality of the equipment.

As stated previously, TRI bases the sensitivity of the gravimetric oil mist and particulate analysis as a function of air volume. The more air that is passed through the sample filter, the more sensitive the reporting limit can be. TRI can have reporting limits ranging from 1 mg/m^3 to as low as 0.0001 mg/m^3. The typical reporting limit for most analysis is 0.1 mg/m^3.

Mold and Bacteria

This report will not go into any great detail concerning mold and bacteria, as it is not a very common problem in the diving industry as a whole. The diving regulatory agencies have no specification limits for mold and bacteria. Mold and bacteria typically require moisture to grow and thrive. As discussed, the diving industry has very restrictive moisture content limits, so therefore mold and bacteria should not realistically pose a safety threat. With that said, TRI does have an "odor test" that needs to be performed as part of the typical compressed air analysis. This is in line with CGA and other regulatory agencies when it comes to compressed breathing gas quality requirements. TRI suggest that if there is ever any pronounced odor that is present in a compressed air system, that the system be tested for mold and bacteria as a part of the standard catalog of analysis. Depending on the species, mold and especially some bacteria can cause serious systemic health issues in the human body. TRI collects mold and bacteria samples on a filter that is fitted in to the flow section, much like the oil mist and particulate sample is collected. A third party organization then collects the filter and test for mold and bacteria colony growth and corresponding identification if applicable.

Conclusions

Diving, whether it is done as recreation or as a profession, can be a dangerous endeavor. One of the most important precautions taken in any dive is to make sure all the equipment is functioning at peak operating capabilities. This will go a long way to ensure that the dive is a successful one. Diving regulatory agencies require that the supplied breathing air being used in this industry is of the highest quality. It is important that the rules and regulations surrounding the breathing air quality be adhered as it is the lifeblood of the sport and/or industry. TRI Air Testing, Inc. is dedicated to the highest standards of scientific analysis. TRI strives to be a reliable partner with all customers to help them maintain high levels of quality in their processes. Please visit www.airtesting.com for further information about TRI and air quality testing.

References

1. ***US Navy Diving Manual, 6th revision,*** United States: US Naval Sea Systems Command 2008.

2. ***CGA G-7.1-2011 Commodity Specification for Air (6th Edition) ,***Compressed Gas Association 2011.

3. ***Bennett and Elliott's Physiology and Medicine of Diving (5th Edition) ,*** A. O Brubakk.; T. S. Neuman 2003. United States: Saunders Ltd.

4. **National Oceanic & Atmospheric Administration (NOAA) – Earth System Research Laboratory (ESRL)- *Various atmospheric study reports on carbon.***

5. ***Carbon Monoxide – Molecule of the Month***, Mike Thompson Winchester College, UK

6. ***Organic Chemistry (5th Edition),*** J. McMurray 2000. Brooks/Cole: Thomson Learning. ISBN 0-495-11837-0

7. ***American Review of Respiratory Disease ,***John M. Clark(1974)

GREENSEA

COMBAT DIVER PROPULSION AUTOMATION AND INTERFACE USABILITY

A Diver-Friendly User-Interface Offering Navigation, Mission Planning, and Automation

Author and Presenter: Ben Kinnaman, CEO and President, Greensea Systems, Inc.

February 15, 2017
2017 Underwater Intervention Conference

INTRODUCTION

Accurate subsea navigation and integrated mission management tools offer the potential of significantly enhancing the job of the modern combat diver while reducing his work load. However, current systems on the market have found limited adoption because they either do not fully deliver the capabilities required by the operational community, or they are difficult to use in a combat scenario. There exists a significant need and opportunity for a compact diver navigation and mission planning tool that can be used by either a swimmer or a team in a propelled diver transportation system.

STIDD Diver Propulsion Device

To address this need, STIDD Systems, Inc. (STIDD) and Greensea Systems, Inc. (Greensea) partnered to develop a new concept for combat swimmer navigation and mission management, RNAV2. It builds on proven systems from both companies and leverages a combined 25 years of product development and testing in this domain. Using Greensea's powerful OPENSEA software operating platform, the RNAV2 provides advanced capabilities for STIDD's Diver Propulsion Device (DPD).

STIDD fielded the DPD in 2001 and has delivered more than 500 units to the international special forces community. The newest DPD model is capable of transporting two divers with full payloads to distances over 10km, at speeds up to 4kts. The DPD is the most popular and widely-used diver propulsion device in the world.

RNAV2 provides an inertial navigation solution, mission planning tools, and automation to make the DPD—and its pilots—more effective, powerful, and accurate. In addition, RNAV2 can be used as a stand-alone swimming system without the DPD. However, to effectively utilize the system, our team realized the greatest challenge was making it easy to use and intuitive. We also understood that if we were successful, RNAV2 would mark a paradigm shift in combat diving operations.

THE CHALLENGE

Fully-integrated navigation and control software packages such as SmartFlight™, found on the popular Teledyne SeaBotix vLBV, have become the standard among EOD and MCM groups utilizing inspection-class ROVs for critical applications. These systems provide inertial navigation, sensor fusion, automation, sonar integration, task automation, and mission planning in a single-operator interface to simplify the

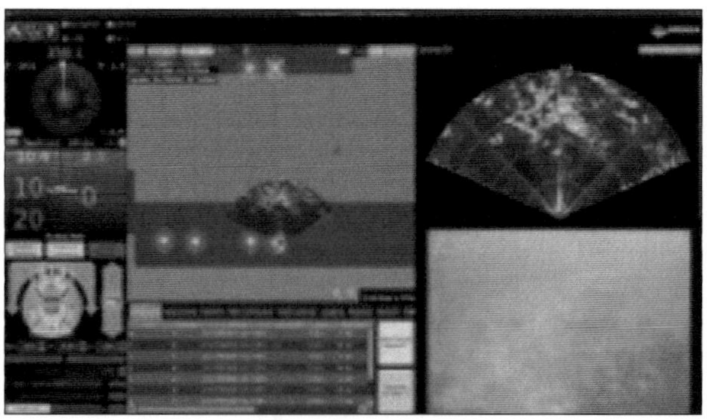

Teledyne SeaBotix SmartFlight Workspace, a Greensea system

operation of the vehicle and increase the effectiveness of the operator. Systems like SmartFlight bring the operator closer to the task at hand by removing the nuances of vehicle operation from the problem. Combat divers can also benefit from the use of this technology, but several operational issues need to be addressed.

With a decade of experience fielding OPENSEA's advanced navigation and mission management systems in more than 600 manned, unmanned, subsea, and surface vehicles, Greensea was not particularly concerned about the technology integration. Further, the DPD is a stable and time-proven platform already embraced by special forces personnel. The barrier to usefulness was not the technology itself so much as the access and utilization of the technology. The real engineering task in this development was making this technology user-friendly in a combat diver application.

This is a familiar theme in robotics, automation, and even personal computers that extends back to the advent of modern computer science. Technology has enabled more powerful sensors that offer new capabilities in smaller sizes. Using these new technologies, we make the machines around us smarter and more capable; however, we cannot simply focus on making a smarter machine. We have to consider our relationship with it and how we as operators will be affected. How will we use it? How will we communicate with it? How will it serve us? This was the critical issue our team encountered when adding intelligence to the DPD – a system well-known and beloved for its simplicity.

Our development team (consisting of both robotics engineers and diver-operators) conversations often mirrored a famous conversation between Artificial Intelligence forefather Marvin Minsky and Doug Engelbart, the creator of the computer mouse, in the 1950s at MIT. As cited by Kevin Kelly in his 1994 book, Out of Control:

> Minsky: "We're going to make machines intelligent. We are going to make them conscious!"
>
> Engelbart: "You're going to do all that for the machines? What are you going to do for the people?"

APPROACH

To develop a user-friendly interface for RNAV2, our engineering team used three primary tactics: create a simple interface, increase operator bandwidth, and test regularly. This approach was supported with meaningful training and education for the operator.

*Provide a simple and task-specific **interface***. Greensea's user interface, commonly used in ROV applications, needed to be reconceived for the diver application. While the task-specific architecture combining navigation, vehicle control, mission planning, payload management, and imaging sensors remained intact, the new implementation recognized that a typical ROV operator sits on a ship in a warm, dry control room, often with support staff – a dramatically different environment than a diver piloting the DPD in zero visibility at 4kts underwater.

Greensea tests the STIDD RNAV2 in Lake Champlain, VT

To provide a similar toolset of capabilities, we completely re-imagined the user experience. We focused on compartmentalizing the operator's tasks and understanding the diver's workflow through a wide variety of mission profiles – from swimming to driving the DPD. Once we had a software architecture that matched the operator's in-water workflow, we designed a fast and simple input methodology that absolutely minimized the diver's physical movements.

We employed a software architecture featuring two sets of stacked screens that match the diver's operation profile – the machine is either working for the operator (operations), or the operator is working on the machine (diagnostics). The operations stack consists of primary operating screens and provides task-specific screens for pre-dive checks, fully-aided subsurface navigation on a chart-plotting screen, limited-aid subsurface navigation in a familiar "Tac-Board" format, and a sonar-only screen. These primary operations screens allow the operator to "swipe left and right" through the stack as his mission progresses, simply keeping only the most task-relevant information on the display. The diagnostic stack of screens is accessed through a workflow outside of the normal operating stack and provides detailed information on the health and status of RNAV2, calibration tools, and an alarm history.

RNAV2 provides two ergonomic handles to keep the unit correctly positioned for swimming or DPD operations. The goal of the input design was to keep the operator's hands comfortably on the RNAV2 handles and literally place all functions under a thumb. The exception to this was the desire to have several mechanical buttons accessing specific task screens and a panic button that could be mapped to muscle memory. RNAV2 provides a three-axis joystick under each thumb and a rocker trigger under the diver's fingers on each hand. The controls on the right hand are the mouse functions and the controls on the left hand are function controls that change with the displayed screen. The diver accesses each task screen through a familiar "swipe-left" or "swipe-right" motion of the left thumb and navigates a simple pop-up menu structure with the left trigger buttons, up to select and down to display task options.

Increase operator bandwidth through automation and autopilots. Once we started testing RNAV2 in open-water with the DPD, we realized we had made the machine more intelligent, but were mindful of Engelbart's motivating question to Minsky: "What are you going to

RNAV2 diver navigation system

do for the people?" Although RNAV2 was easy to use, even its simplicity required too much from the operator while piloting the DPD. To overcome this, we implemented a complete fly-by-wire and automation package to increase the operator's bandwidth by removing the task of manually flying the DPD.

The DPD, like most diver propulsion devices and manned submersibles, uses a very simple directly-actuated control surface system for maneuvering. The diver simply turns the steering wheel to adjust heading and pushes and pulls the yolk to adjust his pitch angle. We maintained this mode of operation, as well as the simplicity and robustness, but added parallel electronic control to the actuators. With this, we implemented a fly-by-wire autopilot mode and a full automation mode for guidance. The fly-by-wire mode allows the diver to maneuver the DPD with only his right thumb joystick by controlling the heading and pitch/depth autopilots. This allows the diver to "set-and-forget" the heading and depth controls while freeing him to focus on other tasks. Further, it gives him the ability to fly the DPD without ever removing his hands from the RNAV controls.

We also added the complete mission automation capability used in Greensea's ROV and AUV software packages. With this system, divers can pre-plan missions fully establishing mission parameters: waypoints, tasks, transit depths, navigational markers, and exclusion zones. Once in the water, the diver enables the autopilot with a single button and RNAV2 executes the pre-planned mission, completely automating the steering and depth control while the DPD transits through the selected route. We also added functions to modify routes, create new routes, go to specific waypoints, and even functions to mouse-click-to-go on the chart, that provide the operator complete control of the DPD through RNAV2.

Test early, test often, and listen carefully. At the onset of this project, we felt the only way to develop an effective user interface for this technology was to intimately link the software development process to the open-water operation of RNAV2. Our development team consisted of engineers and operator-divers, a cliché-ishly comical software development team that drove the rapid success of the program. Further, all of the software developers at Greensea involved in this project were divers and had regular diving responsibilities during the year of testing and

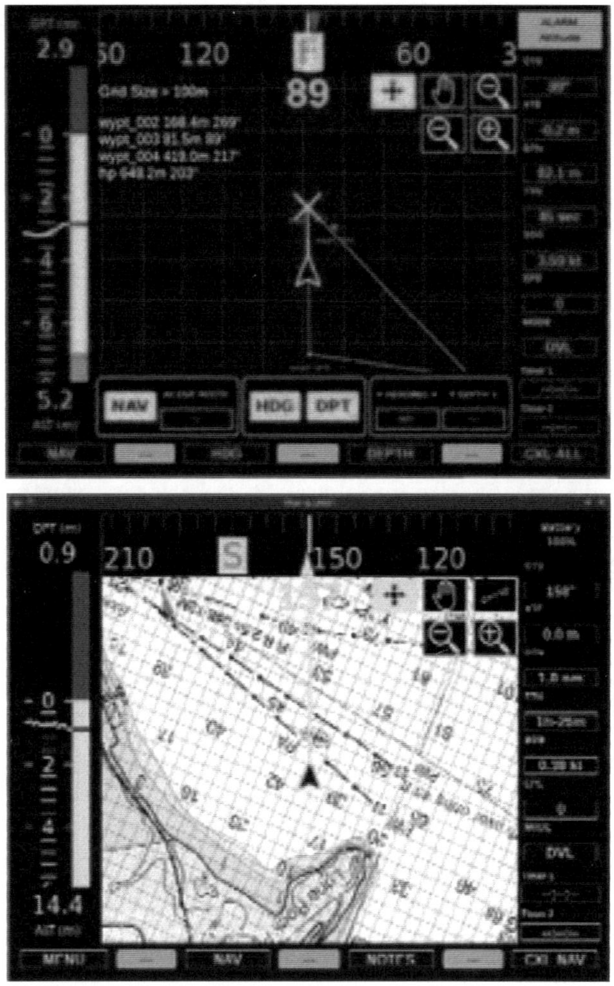

STIDD RNAV2 Workspace user interface

developing RNAV2.

Our team tested almost weekly in open water during the year of development. Our initial testing was in the warm, clear waters of Marathon, Florida, at STIDD's facility, and then moved to the more operationally-realistic, dark and cold waters of Lake Champlain at Greensea's facility in Vermont. Our testing focused on functional verification of the software through module and system testing, but most significantly on the usability of the software. Functionality testing verified the software's adherence to requirements, while usability testing verified the user-experience of operating the software and focused on four main data points: performance, witness data, survey data, and path-to-aptitude.

Understandably, a user interface that is hard to use will adversely impact the overall system performance. As an example of this, consider the text messages we receive today versus the text messages we received when we all carried flip phones. To test operator performance using RNAV2, we assigned divers complex and multi-layered, yet real-world, tasks and evaluated their performance. Examples included finding an object with a sonar while swimming in mid-water, creating new routes while flying a current route, and adjusting payloads or changing positions on the DPD while in transit. We evaluated the effectiveness and usability of the RNAV2 interface based on an operator's performance accomplishing an assigned objective.

Witness data, video of the operator using RNAV2 during a dive, and survey data provided critical feedback on how the operators were using the interface and their experience doing so. During much of our testing, we used a GoPro camera mounted to the diver's head while swimming and a GoPro camera mounted to the DPD to record the diver's use of the software through specific tasks. We also asked each diver to regularly complete a survey implementing a System Usability Scale (SUS) questionnaire that helped us quantify the usability of the interface.

Lastly, the path-to-aptitude was an important piece of test data to determine the usability of the interface. Path-to-aptitude shows how quickly a new operator can become functionally proficient with the software and helps inform us on intuitiveness of the software. A user interface that has a short path-to-aptitude is easy to learn, intuitive, well-liked, and often robust. Our team felt that ensuring the RNAV2 software had a short path-to-aptitude would, not only make RNAV2 easy to use, but also aid a diver in successfully resolving an emergency or stressful situation because it would be intuitive, easily memorized, and clear.

CONCLUSION

The task-based architecture of RNAV2 provides a focused and compartmentalized display of data that is specific to the diver's operational mode. Testers find operating the RNAV2 with dual hand controls intuitive and powerful. Test video at the end of the development period showed divers successfully conducting 3km and 4km submerged complex missions without ever having to remove their hands from RNAV2's hand grips, and having the bandwidth to attend to mission and payload tasks while underway. Survey data from divers shows the Workspace is easy to use and well integrated into the mission workflow.

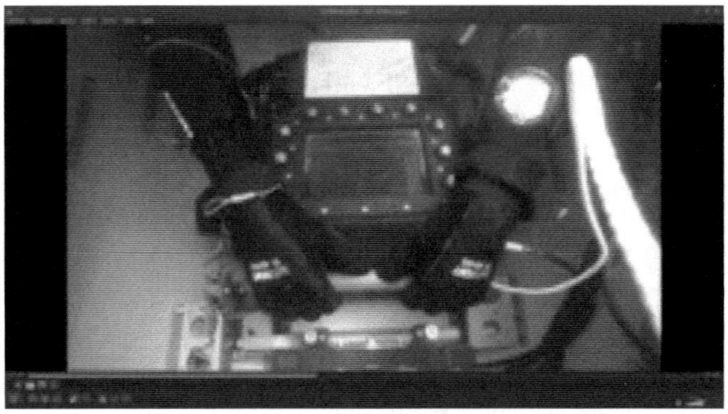

Diver flies the STIDD RNAV2 AP2 hands-free

Many attempts and existing solutions for advanced navigation and payloads for divers have failed, or not reached their full potential, because even the simplest solution added too much workload to the diver. We found that, ironically, through the addition of technology we could reduce the diver's workload. We also found the use of autopilots and route automation in RNAV2 significantly and profoundly decreased the workload on the diver, freeing him to use the navigation system as designed, make use of the sonar and mission planning tools, and focus on the mission objectives. Quantitative test data shows the use of autopilots and route automation in RNAV2 has two other significant benefits: better performance and decreased training requirements. We found that divers could not only follow a mission more exactly and efficiently, they had to train less to do so with autopilots flying the DPD.

A weekly testing schedule for any software package is aggressive, but testing a diver system weekly takes significant commitment. Our team feels the success of RNAV2 is largely due to our commitment to test early, test often, and listen closely to our operators. Test data included quantitative control and navigation data, requirements verification data, quantitative performance data, and user-experience data derived from witness video and direct surveys. This data provided a vital, regular feedback source to our engineering team during development.

The successful progression of man and machine's relationship relies on attention to both the man and the machine. We cannot just make smarter machines without addressing the man and his relationship with the machine. If we provide operators the communication tools needed to work with their smart machine partners, the work they do together will be more effective, more efficient, and more productive. Further, addressing the man-machine relationship is critical to fielding advanced technology in machines. Perhaps no better example of this exists than the effort to put advanced technology in diver systems. During the development of RNAV2, we made the machine smarter and gave the operator the tools he needs to work effectively with this smarter machine. In doing so, we fielded a successful and powerful system to assist the next generation of combat divers.

ABOUT THE AUTHORS

Ben Kinnaman

Ben is the President and CEO of Greensea located in Richmond, Vermont. He has been in the offshore industry almost 20 years as a diver, ROV operator, and engineer. He founded Greensea in 2006.

www.greenseasystems.com

Greensea
10 East Main Street
Richmond, VT 05477
802.434.6080

Bill Hellman

Bill is the Director of Diving Operations at STIDD Systems, Inc. located in Greenport, NY. He is a retired US Navy Seal with 39 years of operational experience in Wet Submersibles and underwater navigation. He continues to work with the military operational communities worldwide to steer STIDD's DPD and RNAV2 products.

Jeff Carr

Jeff is the Diver Training and Operations Manager for STIDD Systems, Inc. He lives in Marathon, Florida and heads STIDD's field office there.

www.stiddmil.com

STIDD Systems, Inc.
220 Carpenter St.
Greenport, NY 11944
631.477.2400

AUTHOR INDEX